The Locavore's Handbook

The Busy Person's Guide to Eating Local on a Budget

LEDA MEREDITH

To my grandmother, Eugenia Kilgore,
close to a century old and still curious about the world.

To buy books in quantity for corporate use
or incentives, call **(800) 962–0973**
or e-mail **premiums@GlobePequot.com.**

Lyons Press is an imprint of Globe Pequot Press.

Quotes on pages 11 and 27 are from *In Defense of Food* by Michael Pollan, © 2008 by Michael Pollan.
Used by permission of The Penguin Press, a division of Penguin Group (USA) Inc.

Text design: Sheryl P. Kober
Project editor: Julie Marsh
Layout artist: Nancy Freeborn

Library of Congress Cataloging-in-Publication Data is available on file.

ISBN 978-0-7627-5548-6

Printed in the United States of America

10 9 8 7 6 5 4 3 2 1

Contents

Acknowledgments

Thanks to:

My main farmers: Ted and Jan Blomgren of Windflower Farm, Nancy and Alan Brown of Lewis Waite Farm, Don Lewis of Wild Hive Farm, and Mary and Bob Pratt of Elihu Farm.

My editors at Globe Pequot Press, Julie Marsh, Paulette Baker, and Heather Carreiro, who was my first contact at GPP.

My agent, Meredith Hays.

Naomi Rosenblatt and Michael Fancello at Heliotrope Books.

Sandor Ellix Katz for the eloquent introduction.

Liz Neves and Kendall Cornell for letting me include some of their personal stories about local foods.

Tom Twente, Judy Janda, and everyone at the Park Slope CSA. Thanks, Judy, for the tips on how to prioritize the vegetables in a share so that they don't spoil before you get to them.

Colin Beavan for the no impact experiment.

My students at Adelphi University, the New York Botanical Garden, and the Brooklyn Botanic Garden.

Penelope Coberly, Kelly Johnson, Francis Patrelle, Jennifer Muller, Julie Voshell, Bill Hedberg, and Anton Wilson for ongoing support.

Foreword

The Locavore's Handbook is a practical guide for people who wish to reclaim the food they eat. *Locavore* is a new word describing local eating, which has been adopted by a cutting-edge social movement. But the locavore concept is nothing new. Not necessarily 100 percent local, but predominantly local is the way almost all people ate until the past few generations. It is mind-boggling how much critical cultural information we have lost in just those few generations. For most of us, it is not at all straightforward how to restore locavore ethics into our diets and our lives. With this book, Leda Meredith shows us where to start.

Food has rhythm. It is never created in a continuous even flow. The passage of seasons dictates what foods are available when. This creates rhythms of abundance and scarcity, an endless ebb and flow, and our ingenuity and effectiveness at preserving the abundance determines how we eat during the periods of relative scarcity. Techniques for saving seeds, rising loaves, fermenting, drying, and otherwise preserving food are all about timing; they too are rhythms. As are the repetitive activities of digging, hoeing, seeding, cultivation, harvest, and processing. All life marches to the beat of its nutritional sources, whatever they may be. Except, perhaps, us.

We can shop twenty-four hours a day and purchase food without regard to the season. Twenty-first-century humans generally regard this as evidence of progress. We believe that we have been liberated from toil by not having to produce or procure our own food. And in some ways we certainly have. Specialization, globalization, all-season produce, and one-stop shopping provide freedom, leisure, convenience, and security, at least to some segments of society. But at what cost?

Unfortunately, the intensification of food production that has removed this formerly universal activity from most peoples' lives is completely unsustainable. Chemical-based monoculture erodes and pollutes soils, depletes and pollutes water resources, produces less nutritious and increasingly toxic food, and is heavily reliant on rapidly diminishing fossil fuel resources. Concentration of production and distribution is making our food supplies more vulnerable to contamination and, at the same time, making sources of contamination more difficult to trace. Transporting most of the food we eat thousands of miles from farm to plate is a major contributor to our carbon footprints and climate change. The mass production model, as it has been applied to food, is not sustainable. It is diminishing our health and destroying the Earth. We have an urgent imperative to reclaim our food and reconnect to the important rhythms of feeding ourselves.

Some organisms have their food procurement strategies hard-wired into their genes. Humans, remarkable in our ability to adapt to different environments with radically different food resources, rely upon cultural information, passed down from generation to generation. The problem with having been severed from essential rhythms for feeding ourselves is that most of us no longer possess that cultural information. We do not have an inbred ability to recognize the rhythms of food in our environment or to syncopate our participation in those rhythms.

With *The Locavore's Handbook,* Leda Meredith is helping to provide us with some of the missing cultural information we need to reclaim our food. This book guides the reader through an understanding of the seasonality of foods, basic techniques for preservation, finding local producers, urban gardening, and foraging. You do not have to become 100 percent locavore and renounce food from beyond 250 miles to use this book. For most of us, change is incremental. But now is the time to start, and Leda gently coaxes and encourages her readers to take small steps toward feeding ourselves locally and participating more directly in the process of feeding ourselves.

—Sandor Ellix Katz

Author of *Wild Fermentation: The Flavor, Nutrition, and Craft of Live-Culture Foods* and *The Revolution Will Not Be Microwaved: Inside America's Underground Food Movements*

1

My Year as a Locavore

From August 2007 until August 2008, I ate almost exclusively foods grown or raised within a 250-mile radius of my apartment in Brooklyn, New York.

I'm not expecting you to do that.

But that yearlong local eating challenge taught me a lot about how to eat local while living a busy, urban life. It also changed me in ways I didn't anticipate when I started the experiment. In fact, my diet is still 80 percent local.

If you think my 250-mile diet was extreme, you're right. I'm not sharing what I learned during "The 250" and after so that you will undertake something so strict for yourself. My goal here is to make it easier for you to incorporate local foods into your meals to whatever extent works for you. It doesn't have to be all or nothing—even a small switch toward a local diet has a major impact. Here's an encouraging fact: "Buying 25 percent of your groceries from local farmers for a year lowers your carbon footprint by 225 pounds—even more than recycling glass, plastic, and cans!"(*Eating Well Magazine,* February 2009)

Why I Became a Locavore

I started The 250 because I wanted to reduce my impact on the environment. That declaration is about as dry as unbuttered toast. It doesn't begin to describe how I felt just before I began The 250—or what motivates me to continue to eat local now. I wasn't trying to be "green" in a trendy way or to assuage eco-guilt. I felt bombarded by each news story of the latest environmental catastrophe that was being intensified by human activity: climate change accelerating at a much faster rate than predicted by scientists just a few years ago; topsoil in the Midwest's farm belt so depleted and polluted that within a generation we may no longer be able to produce safe food there.

So much of the information coming at me was contradictory and confusing. I couldn't reconcile the food labels and magazine articles incessantly urging me to eat more fish for my health's sake with the equally prolific warnings about how many species of fish are contaminated with heavy metals and being overfished to the point of near extinction. Worldwide food shortages and escalating food prices were on prime-time TV, followed thirty seconds later by stories about the obesity epidemic in our country. The United States Department of Agriculture (USDA) came out with a new pyramid of dietary guidelines suggesting the same thing my mother had—"eat your vegetables"—but salmonella-infected spinach was also in the news.

The gloom-and-doom environmental news reports were overwhelming. What could I do? I dutifully put my bottles and cans into the recycling bin in front of my apartment building. I switched to energy-conserving lightbulbs. But those small efforts hardly seemed enough to make a dent in the disasters my species was inflicting not only on other species but also on its own future, on *me*.

One small story buried in an online newspaper report haunted me. I read about children who will never see a firefly because that species is threatened by pesticide spraying aimed at other insects, much of that spraying done on behalf of industrial agriculture. I remembered the sparkling clouds of fireflies I used to see in my garden and realized that, yes, in recent years there have been very few.

Climate change, global warming, environmental issues—whatever label you stick on it, time is running out for us to make substantial lifestyle changes that might ensure a verdant world to live in for generations ahead.

I fantasized for a while about getting out of the city and living off-grid somewhere with clean waters and unpolluted air and soil, a place and a lifestyle that would cause no harm and where I would be untouched by environmental crises and human irresponsibility. Of course no such place exists today. And even if it did, when I was honest with myself, I had to admit that I had neither the financial means nor the deep-down desire to quit my city life. I decided it wasn't a bad thing to keep living in the city. More than half of our species now lives in urban areas—for the first time in human history. If we are going to successfully alter our lifestyles so that this planet can continue to sustain us, much of that change is going to have to be accomplished by city and suburban people. Like me.

Starting Small

At first, changing what I ate was a small change that didn't mean more to me than taking out that recycling or changing to those energy-saving lightbulbs. I knew that conventional agriculture dumps tons of pesticides into the ground, polluting both soil and groundwater. Knowing that organic farms don't use all those pesticides, I spent the extra to buy organic whenever I could afford it.

Gradually I became aware that the impact of my food choices went far beyond the organic or nonorganic issue (more on local versus organic in chapter 2, How Can Eating Great Food Save the World?). I read that in the United States, food travels an average of 1,500 miles from farm to plate. Here in the Northeast it's often closer to 3,000 food miles for produce shipped from California and Mexico. That's a lot of fuel-burning miles. I also learned about the struggles of local farmers and how the gradual disappearance of small family farms over the past few decades has had a devastating effect on local economies and landscapes. I started to see that what I ate for lunch could make a real difference in the world.

When questioned by city people about what they can possibly do about the decline of American farming, Wendell Berry, farmer and author of *Bringing It to the Table: On Farming and Food,* has answered that they should eat responsibly. What does eating responsibly mean? Is it buying local? Buying organic? Swigging coconut water for its health benefits or eschewing white rice in favor of brown? And does that matter when rice isn't grown anywhere within a few hundred miles of me (never mind the coconuts)?

Over the course of several years, my eating patterns gradually shifted away from the corner deli and the salad bar at Whole Foods. I started gardening and growing food on a roof, a fire escape, and eventually in a community garden plot. I discovered that the farmers' markets here in New York City are just as colorful and fun to shop as the ones I'd enjoyed when traveling overseas. When I found out that there was a community-supported agriculture (CSA) farm-share program in my neighborhood, I signed up.

That was my involvement in the burgeoning local foods movement before The 250. It was fairly extensive (signing up for a CSA share is a real commitment), but I still wanted to do something more.

I was inspired to do The 250 when I read about *locavores* who'd done their own 250-mile, 100-mile, or even 50-mile local eating challenges. (Definitions of "locavore," the *American Oxford Dictionary*'s 2006 Word of the Year, vary. Basically the term describes persons who source their food within a specific radius from where they live.)

I was fascinated by Gary Nabhan's *Coming Home to Eat,* Michael Pollan's *The Omnivore's Dilemma,* Barbara Kingsolver's *Animal, Vegetable, Miracle,* and other books by people who translated their passionate desire to live in balance with the natural world into a year of eating within a strictly defined radius of food sources. As I read up on why eating local mattered, I learned that there are even more reasons for eating from the region you live in, reasons that cross international, political, and economic borders. I learned about the damage U.S. agribusiness subsidies and trade policies do to small farmers in other nations and how the industrial food system is the direct cause of health scares including swine flu, as well as the unholy combination of obesity and malnutrition that has created a surge in diet-related diseases such as diabetes.

I was inspired by what I'd read about others who'd done a local foods challenge year (see the Useful Resources appendix for a full list of recommended books on the subject), and I thought I was a good candidate to make it through such a year successfully. I already had a few skills that would make it easier: I loved to cook, I'd already done some food preservation as a hobby, I gardened, and I'd been a wild edible plants enthusiast since I was a kid.

Could I Do It?

It seemed as though I would be a sure bet for a locavorian year, but nonetheless I had a few reservations: I found it discouraging that every one of those authors whose books I'd read had advantages over me when it came to eating foods raised within their foodshed. They had gardens so large they amounted to small farms, had much more money than I did, were retired from work and had lots of spare time for food projects, or lived in a gentler climate with a longer agricultural season. Could it be done on a budget, living in my tiny one-bedroom apartment, in a climate with cold winters, and despite my super-hectic work schedule?

If I could do it, then I might be able to inspire others to do the same—or at least to get some of their groceries from local sources. I knew that many of the friends

I hoped to inspire didn't share my enthusiasm for cookery and plants, but I hoped having those skills would make it easier for me to be a trailblazer and motivate people who heard about my experiment to make at least some shifts toward eating a local foods diet. My example might encourage people like my friend Sarah. She had read all the same books about the local food movement that I had, and understood all the reasons it makes sense to eat local foods, but balked because "let's face it, I can't do that. I live in the city and have a full-time life."

I have a full-time city life too, I thought.

I wasn't expecting Sarah to take it to the extreme of a 250-mile diet, but I thought she might be able to do more than she was—and that maybe by doing The 250, I'd be able to show her and others that it wasn't as impossible as they believed.

The day before I started The 250, I sat down with a map of the Northeast and measured out 250 miles from New York City in every direction. I drew the circumference on the map and cut out that piece of the United States. I stared at that circle of paper for a long time thinking, *This is my world for the next year.* Then I stuck the map under a magnet on my refrigerator and started to think about what I would (and wouldn't) get to eat the next day.

In addition to doing my bit to save the world, I had a purely selfish reason for eating local. Even before The 250 I knew that locally grown food was fresher, picked perfectly in season, and by far more delicious than anything on the supermarket shelf—more delicious even than the organic stuff shipped from across the country. That's why so many restaurant chefs are champions of local, seasonal ingredients and have done a great deal to publicize and support local farmers and the local food movement. Food choices are unique in having an impact on the environment *and* offering an immediate sensory reward when you do right. Other "green" choices just don't have the same payback. I can bring my own reusable water bottle with me everywhere rather than tossing out plastic water bottles; but although that is an important eco-friendly choice, it doesn't do much for how I feel beyond a mental pat on my back for doing the right thing.

But food is different. If I eat that environmentally expensive out-of-season strawberry, I am disappointed because it just doesn't taste like much. But if I wait for the local strawberries to be perfectly ripe—the strawberries that didn't burn as much fuel to get to me—they burst on my taste buds with sweet, juicy gratification. Saving the world tastes good.

Making It Personal

Everything that came into my kitchen during The 250 had to pass through a kind of border control. I looked up the farm each item came from online to see if it was within my 250-mile radius. I started to have a more detailed knowledge of the region than I'd ever had before, even though I'd lived in New York City for close to thirty years. I came to know exactly where the Finger Lakes are upstate and precisely where the North Fork part of Long Island is because my favorite local wines come from those areas. I learned that my favorite apples come from the Hudson Valley and that the first tomatoes of summer come to me from South Jersey. I started to feel a sense of place, of belonging where I live, and I started taking pride as well as pleasure in my region's abundant food.

I also came to care about the farmers and the land that feeds me in a personal way. Some of those farmers have become friends. Since The 250, each time I set a plate on my kitchen table, I am aware that those are Ted's green beans and potatoes, Nancy's pork, George's fennel. There are names and faces and memories attached to my food now. At the 2008 Brooklyn Food Conference, Dan Barber, the chef for Blue Hill restaurant, said that none of the evils of industrial agriculture could have gotten so extreme if there hadn't first been a separation between community and food supply. A local foods diet rejoins farmer to eater, region to inhabitant. On a personal level, this translates into moments like the one when my grain farmer and miller, Don Lewis of Wild Hive Farm, steps out from behind his stall at the green-market to give me a hug, I feel honored and grateful that I know him personally. It's an immeasurable ingredient, but I'm sure the bread I make with his flour tastes better for it.

What Here Tastes Like

My original title for this book was *What Here Tastes Like* (the subtitle about budgets and busy-ness was the same). My publisher nixed that for the reason that nobody would have a clue what it meant. But there is still something about the concept of "what here tastes like" that intrigues me. Other places have tastes associated with them. If I say "Italian food" to you, I bet something comes to mind immediately. Likewise, if I say "Thai food," "Mexican," or "Japanese," your memory of eating those foods may be vivid enough to make you salivate (assuming you like that particular cuisine). Those are place names richly associated with the flavors, textures, and aromas connected to their cultures. Why shouldn't our place have a taste? You could say New York City tastes like bagels, or some other food that has become ubiquitous and traces its history to the diverse waves of immigration that came through our port. Those foods do belong in the "what here tastes like" category, as does the magnificent array of cuisines from every part of the world that reflect how international our city is. But those tastes do not offer a sensory hold on the agriculture of our region, the particular mineral makeup of our soil—what the French call *terroir*. I am still discovering what here tastes like.

More Than Expected

So those were the reasons I began The 250: I wanted to do something to help the environment, I wanted to support small local farms, I wanted to eat great food, and I wanted to prove that it was possible to eat an almost exclusively local foods diet without owning a farm or moving to the West Coast. What I didn't expect was to come out of the year with a very different view of the city and region I live in, fifteen pounds thinner, feeling more energetic than I ever had before, and having both lost my craving for nonlocal foods and found a greater appreciation of them at the same time.

During The 250 I had many conversations with people who were intrigued by the local foods challenge I'd set for myself and applauded my reasons for doing it but had lots of questions and many excuses for not doing the same. They thought a local foods diet would cost too much, take too much time, lack variety, and be inconvenient. I had those same concerns and set out to see if they were as daunting as they seemed. What I learned is that a local diet *can be* expensive and time-consuming, but it *doesn't have to be* (see chapter 6, The Cost Factor, and chapter 7, The Convenience Factor). It did require more planning ahead than I was used to, but once I'd mastered that, eating local became as convenient as taking a jar down from my kitchen cabinet and calling the (delicious) contents dinner on evenings when I was too tired to cook. As for variety, eating completely within the agricultural calendar actually restored excitement to my menus (see chapter 4, Eating with the Seasons). I found ways to stay within my budget and to fit in time for shopping, cooking, and preserving local ingredients for winter without straining my already packed schedule.

It took a while to find that balance, though. During the first few months of The 250, I was still juggling the cost, time, and space issues. At first I spent a lot of time online trying to find out if certain ingredients that I hadn't seen at the greenmarkets were even available from local sources, as well as keeping track of which days I could get what from the markets. Tracking down certain ingredients such as flour from local sources proved to be a challenge. Until I did find a source, I did without bread or pasta. But the locavore lifestyle got easier and easier as the year went along.

With help from Local Fork, a Web site devoted to helping people in the New York City area eat local, sustainably grown food, I shared what I was learning in an online directory called The Locavore's Guide to New York City (NYC). I was doing the research and legwork to make other would-be locavores' lives easier, I hoped. It was encouraging when I started getting e-mail from people letting me know about the great local wine they'd just tried that I might like, the no-cook meal they put together from farmers' market finds, or the restaurant they'd discovered that featured local ingredients. It was delightful that even though I didn't insist my friends stick to local ingredients when they invited me to dinner, they often made a partially local feast for me anyway.

Being a locavore in the twenty-first century isn't about the back-to-the-land movement of my parents' generation or being a slave to the kitchen as my grandmother had to be. It's about a partnership with the farmers and the ecosystems that feed me. I'm grateful that I don't have to spend all of my time ensuring that there is food on my table. The farmers who do that for me are my heroes. But they need me as much as I need them.

Much of what people think of as New York City was invisible to me during The 250. The coffee shop wasn't an option, and after a while I stopped noticing it as I walked past. I also failed to notice a new restaurant that opened up less than a block away from me until a friend pointed it out—it simply didn't pertain to me. Instead my mental map of the city became centered on the greenmarkets—e.g., Tuesday is the Borough Hall market in Brooklyn, where the mushroom guy also sells popcorn. My mental geography reversed itself. Unlike that famous picture in which New York City takes up most of the map and there's just a little space allotted to everywhere beyond it, I started thinking of myself as living in the center of a circle that extended well beyond the city's borders. Although I still lived in the city, my sense of home now included Ted's farm 180 miles northwest; the fishing coast of Montauk, Long Island, to the east; Kernen Farm in South Jersey; the maple syrup sugar shacks in Massachusetts; and every other bit of land that was literally keeping me alive. Instead of narrowing my sense of where I live, The 250 actually expanded it. As a result, I still live with a more detailed awareness of the land and sea surrounding New York City that provide my sustenance.

A Few Exemptions

Despite becoming somewhat blind to the restaurants and convenience food offerings around me, I am not anti-restaurant or anti-deli, or anti–international food trade. I am well aware that one of the advantages of living in a metropolis like New York City is the wealth of food choices available. Even during The 250 I wasn't a complete purist. I had ten rules that granted me exceptions to the prime directive of eating only what was produced within 250 miles. One of my rules was that I could eat out twice a month with friends or at their homes. This rule was really to make things easier on them. I didn't want them to stop inviting me just because it was too much of a hassle for them to come up with a 100 percent local meal.

I also granted myself three ingredient exemptions: salt, olive oil, and coffee. The first two were because I couldn't imagine cooking without them. The coffee, well, I don't really have an excuse for that. Most locavores have similar lists of exemptions and rules. The point is to set rules that you can actually live with. When I made up mine, I tried to choose guidelines that I thought I'd be able to stick to. I didn't want to be so impossibly strict that I'd give up early or feel the urge to "cheat."

You'll find your own can't-live-without-it exemptions on the locavore path. Can't imagine cooking without lemons and other citrus? Fine, that becomes one of your own personal "rules." (By the way, citrus, though it doesn't grow in our climate, is in season where it *does* grow from late fall through early winter. It is much cheaper then—think six lemons for a dollar versus three for the same—and much juicier.)

The olive oil and salt exemptions that I gave myself during The 250 were not special-occasion treats—I cooked with them every day. Still seeking to minimize my carbon footprint even when it came to these exemptions, I did some research on the salt and found that I could get sea salt from Maine—well outside my 250-mile radius but as close as I could get (that is still the salt I use). I bought organically shade-grown, fair-trade coffee as a best-I-can-do-without-giving-it-up option.

The oil was more complicated. In theory you should be able to get vegetable oils from the northeastern United States where I live. Olive trees can't survive our winters, but there are walnut trees that do, and you can grow sunflowers here. Those could be turned into cooking oil. Except no one within my 250-mile radius *is* turning them into cooking oil. So if I was going to be a purist, my options for what to heat up in the omelet pan or sauté my onions in would have been local butter (divine, by the way) or rendered animal fat. Buying organic olive oil was my compromise.

Going Hard-Core

Six months into The 250 I decided to go hard-core. I limited my olive oil exemption to a few times a week. I started saving the rendered fat from my bacon and duck. (You've gathered by now that I'm not a vegetarian. If you are, don't vanish. There's a lot in this book for you too.) For weeks I saved those rendered fats but didn't use them. I was too scared by all the "saturated fat is evil" claims that I'd grown up with, even once I learned that the scientific data on them isn't nearly as incriminating as I'd believed. (See Michael Pollan's *In Defense of Food* and Sally Fallon's *Nourishing Traditions* if this piques your curiosity.) Eventually I got over my squeamishness and started cooking with butter and those rendered animal fats I'd been saving. I even popped my locally grown popcorn in fat from locally raised ducks. I also started limiting my food source radius to just 100 miles twice a week, just to see if I could. I didn't starve. I felt great. I had energy to burn—and I lost fifteen pounds without trying.

Don't get me wrong: I'm not promoting some trendy new rendered duck fat diet for weight loss and increased energy. Many locavores are also vegetarians. I'm just telling you what happened to me.

Another choice I made when planning for The 250 was that I decided to stop taking any vitamins or other nutritional supplements. I wanted to prove that my body could get everything it needed just from the local foods I was eating. This concerned my friends, two of whom even asked me whether I'd get scurvy in winter because of the lack of fresh fruit. Scurvy! Now there's a word you don't hear often outside of history class. I didn't get scurvy, but I did do some research to figure out where I would get my vitamin C from in winter. I learned that the fruit I had stockpiled in the freezer would take care of that. Interestingly, the lacto-fermented foods like sauerkraut I'd been making were also high in vitamin C—so much so that Captain Cook included rations of sauerkraut to keep his sailors from getting scurvy. (For more on lacto-fermented foods, see chapter 9, Simple Food Preservation.)

I have a few thoughts on why I felt healthier and healthier as The 250 progressed. A local, seasonal diet includes more fruits and vegetables than most people eat. When I picked up my CSA share each week, I couldn't stand to see any of that beautiful organic produce go to waste. So I either ate it all or preserved some of it for winter eating. There was always a rainbow of vegetable colors on my dinner plate, which meant that I was following both the USDA's and my mom's nutritional advice.

More than that, each fruit and vegetable I was eating packed more nutrition than its supermarket counterpart. According to the USDA, the vitamin and mineral content of many crops has dropped 15 to 38 percent since 1940. As Michael Pollan writes,

> *Clearly the achievements of industrial agriculture have come at a cost. It can produce a great many more calories per acre, but each of those calories may supply less nutrition than it formerly did . . . you now have to eat three apples to get the same amount of iron as you would have gotten from a single 1940 apple.*

Fortunately, with food grown by small organic farms, you get a dramatically different picture. They are grown in healthier soils than the chemically drenched monoculture farms of industrial agriculture. By getting my food from local organic farms, I get to eat the equivalent of that 1940 apple. Maybe that was why, starting with The 250, I've found myself satisfied by smaller meals. My body gets what it needs from my meals, and the size of my appetite has adjusted to that lovely fact. Or maybe it's all the fiber in those fruits and vegetables and locally grown grains that's filling me up on less. Whatever it is, I'm not complaining.

Another one of my challenges during The 250 was lack of space, something many city-dwellers are familiar with. I don't have a pantry to store winter supplies; nor do I have room for a spare freezer or anything like that. When I was on *The Martha*

11

Stewart Show, she asked me if I had a root cellar. When I replied that I didn't, she exclaimed, "Oh, you must have a root cellar!" Right, I thought, I'll mention that to my landlord right away. I came up with some unusual ways to store my preserved foods without feeling like my local diet was crowding me right out of my one-bedroom apartment (see chapter 12, The Space-Challenged Locavore).

Beyond The 250

As The 250 neared its end, many friends asked me how I was going to celebrate. Would I go straight for the chocolate, the cashews, the citrus, or the avocados? One couple even offered to take me out for Kobe beef imported from Japan. But I found that I wasn't craving anything in particular. The food I was eating was so good that although The 250 was technically over, for the most part I just kept going with it.

Barbara Kingsolver had a similar experience at the end of her own 250-mile year. When she later checked her journal to see what they had eaten on the day after the year ended, she found that she hadn't even bothered making an entry because they just kept eating as they'd been doing. She also writes that once in a while, they may enjoy a nonlocal wine or other imported treat. Now that The 250 is over, I do too— once in a while.

The key is in that phrase "once in a while." Eating more imported foods than local foods is not sustainable for our planet or our farming communities, but let's be realistic. People are going to indulge once in a while, myself included. Avocados in New York, maple syrup in Los Angeles—such things can be enjoyed as occasional treats but are environmentally destructive when they become an everyday way of life. Since The 250 ended, I eat nonlocal foods occasionally, but not at every meal and not every day. My daily fare is still almost all locally grown and raised. I don't crave imported foods or feel deprived because I only eat them on rare occasions. On the contrary, I enjoy them all the more when I do have them. I'm reminded of an eighty-five-year-old man I know who says that when he was growing up, the only time he ever ate an orange was when he got one in his Christmas stocking. He still remembers how colorful and dripping with sweetness that orange was. In fact, his memory of those once-a-year oranges seems more vivid than his experience of the orange juice he now drinks every day.

The morning after The 250 ended I wrote on my blog, Leda's Urban Homestead:

I do feel a sense of accomplishment for having done the whole year, but all morning I've been hit by waves of melancholy and part of me is sad that it's over. I think I've figured out what that is about.

In many ways, The 250 simplified my life. I stuck to the rules I'd set for myself, and that meant that if it wasn't a local ingredient, I didn't eat it. I'd stand in line at the Park Slope Food Co-op, and the people pulling things off the shelves that had lots of packaging and long ingredient lists seemed far removed from my life.

But now I've rejoined the world of choice. I went to the co-op this morning and walked through the aisles knowing that technically I could buy anything I wanted to. Would I dare to eat a mango? Would it be worth the fuel burned and the most likely underpaid labor somewhere thousands of miles away? Did I even want to? No, not today.

My commitment to eating locally grown food remains strong. But now there are choices I'll have to make every day that I didn't even have to think about during The 250, and that is where the melancholy comes in. In a way, it will take more motivation to eat locally now than it did when I had a list of rules to follow.

Once I got over that post-250 melancholy, though, I discovered that being a locavore had become my normal way of life. Dinner conversations at friends' homes or at mine often found their way back to what I had discovered about eating local. During one of those conversations, I realized that since the cells of our bodies completely replace themselves every seven years, if I kept up with my local foods diet, soon I would literally be made out of the place where I live. But the questions I had had myself when I started about how to eat local on a budget in the midst of a busy urban life were what came up most often.

This book is the result of all those exchanges during The 250 and after. I hope what I have learned will make it easier for you to help save the planet by eating great food.

2
How Can Eating Great Food
Save the World?

On the one hand, that's a scary thought—that the way we live can't last—but we could consider it an opportunity. When you look at it that way, the questions change. Are there more enduring satisfactions we could find without harming the planet? Is there a better life we can have? What, in other words, should be our new tradition?

—Colin Beavan, *No Impact Man*

Wendell Berry famously wrote that eating is an agricultural act. It is also an environmental and political, almost revolutionary act nowadays to choose foods that are outside the industrial food chain.

But before we get into the saving the world part, let's talk about the food.

Local Food Tastes Better and Is Better

For over half a century, conventional agriculture has chosen food varieties for shipping durability and shelf life, not taste or nutritional value.

The food you get at the farmers' market or from your garden or CSA spent at most a couple of days, maybe just moments, between its source and arriving at your mouth. Instead of two or three varieties of lettuce or apples or other food, there are hundreds of each available to you, varieties that were developed because they taste fantastic. This food was picked at its peak with no thought of shelf life.

Besides the fantastic tastes and textures and colors, freshly picked food packs more of a nutritional punch. The vitamin content of fruits and vegetables starts to diminish the second they are picked, so the less time your meal spent getting to you, the better. The berries you get from your CSA farmer or at the farmers' market contain more vitamins than the ones that spent several days on a truck and then sat on your supermarket shelf for a few days more. As for minerals, those come into plants from the soil. Spinach grown in rich, organic soil has more of the iron Popeye's favorite food is famous for than spinach grown in nutrient-depleted chemically farmed soil.

The Illusion of Abundance

When I moved to New York City thirty years ago, there were two kinds of lettuce at the supermarket: iceberg and romaine. There were also usually only two kinds of apples: Red Delicious and Granny Smith. It is definitely easier to find varied produce now than it was in 1979.

But despite the apparent abundance in our stores, the hard fact is that 96 percent of our commercial varieties of fruits and vegetables have gone extinct during the past century. There may be more varieties available now than there were thirty years ago, but there are far fewer than there were seventy years ago. The reason is that sturdiness and shelf life have been conventional farming's priority. A variety of lettuce that was tender and delicious had no place in commercial trade if it couldn't also stand up to being shipped a few thousand miles and then sitting on a shelf for days before it was eaten. So farmers came to prefer the varieties of fruits and vegetables that *could* stand up to that kind of treatment.

Uniformity was and still is another trait preferred by conventional agriculture. When those potatoes go through the shredder en route to being fries at Burger King, the machinery works better if they're all approximately the same size. Not necessarily more tasty or nutritious or interesting, but better for the machines.

For the past few decades, a crop's uniformity, shipping ease, and shelf life have become farmers' top priorities. And because that's what they grew, that's what we ate.

Why Eating As Usual Isn't an Option

One objection to local eating I sometimes hear is that communities have always traded goods to obtain food they couldn't produce themselves. True, but what they were trading and how they moved it around had quite a different impact in earlier times. Travel by wind-powered (and sometimes slave-powered) ships had minimal environmental impact—if environmental impact was even something people considered then, which it wasn't. Overland routes by horse or camel, ditto. Only in the past few generations has it become possible to ship perishable foods quickly enough by train or truck, and eventually with enough refrigeration, so that even tender items like salad greens could arrive at a destination thousands of miles away in edible shape.

Interestingly, many of the diseases we accept as tragic givens nowadays were extremely rare before the beginnings of industrial agriculture, including diabetes and many types of heart disease and diet-related cancers.

The first pesticides used as part of what was then called the "Green Revolution" were made out of the surplus of chemicals left over after World War II. These included Zykon-B, which had been used for mass killings in Nazi death chambers. An industry devoted to producing these chemicals was suddenly about to lose a major customer, the military, and it sought alternative uses for its products even before the war was over. It turned out those chemicals were useful for killing insects and weeds, and the modern agricultural system was born. The hope was that the combination of chemical fertilizers and pesticides would reduce labor and increase crop yields. That proved to be the case. Unfortunately, the benefits came at a huge cost to our environment and our health.

17

Do You Know . . .

1. *the average distance a farm truck can get on one tank of gas?*

2. *how much fertile topsoil in the Midwest has been destroyed by industrial agriculture?*

3. *how much pesticide use has increased since 1996?*

4. *how many gallons of fuel are used for industrial agriculture in the United States?*

(ANSWERS ON THE NEXT PAGE)

Agriculture is only one piece in the jigsaw puzzle of human activities affecting the environment, but it is a big one. One in every five gallons of fuel in the United States is used for industrial agriculture. The majority of the farmland in this country is covered in a blanket of pesticide dust and produces crops never intended to be eaten directly as human food. Most corn, for example, is turned into animal feed (never mind that it is fed to animals that get sick on a corn diet) or into products such as high fructose corn syrup. Chemical farming is destroying fertile topsoil at thirty times the natural rate. Waterways are polluted with the chemical runoff from industrial farms. Food-related illnesses such as mad cow disease, swine flu, and *E. coli* and salmonella outbreaks are in some cases caused by or exacerbated by the new, now conventional farming system.

The current industrial food system is a failed experiment, but it is only a blip on the tens of thousands of years of human history. We've only actually been eating in this petroleum and chemically dependent way for about sixty years. Okay, so *that* didn't work.

The good news is that those of us who care about where our food comes from and how it was grown *are* making a difference. Customer demand for fresh, local food has caused the number of farmers' markets nationwide to jump from 1,755 in 1994 to 4,685 in 2008. The organization Local Harvest reports on its Web site that the number of farms participating in CSA programs has exploded from just 50 in 1990 to 2,700 today. The number is even higher according to the USDA's 2007 census. They list the number of CSA farms as 12,549.

But even though "local" has become a food buzzword nowadays, as a locavore I am still often regarded strangely. When I'm on the subway and have a snack attack,

(ANSWERS TO QUESTIONS ON PREVIOUS PAGE)

1. The average farm truck can travel approximately 250 miles on one tank of gas.
2. Fifty percent of the fertile topsoil in the farmlands of the Midwest has been lost in the past one hundred years due to industrial farming practices that create soil erosion. This rate of depletion is thirty times greater than the natural rate.
3. The increase in pesticide use since 1996 amounts to fifty million pounds.
4. One in every five gallons of fuel in the United States is used for industrial agriculture.

I pull out a small container of my home-dried apples. People do a double take. But right across from me there is someone digging into a bag of Mickey D's fries, and nobody even glances at that person. A friend comes over with bread purchased at a greenmarket and I ask if she knows whether it was made with locally grown wheat (some of the baked goods at the greenmarkets are, some aren't). The look on her face makes it clear that it never occurred to her to ask, and I feel bad that I might have made her feel guilty about her well-intentioned guest offering. I eat the bread. My commitment to local foods is sometimes trumped by a greater loyalty to my friends' generosity.

Most people continue to eat as if it is forty years ago and we are ignorant about the impact of our food choices. Because we grew up eating that way and everyone around us eats that way, it seems normal and excusable. People who choose to eat differently, who think twice about where their food comes from and what effect it has on the world, are still looked at as being a little crazy; admirable in many ways, but still crazy. The following fable describes the situation succinctly.

THE KING AND THE WELL

There was a town that had two wells. One supplied the castle at the top of the hill; the other serviced the villagers below. The village well became contaminated with something that made the villagers go mad. They didn't realize they'd gone crazy, because everyone around them had too. It seemed normal, since everyone was in the same condition. Meanwhile, the king in the castle on the top of the hill was drinking water from his well, which was not contaminated. One day he went down into the village to see his people and to accompany his chef on a shopping trip. He thought it would be a fine way to connect with his constituency.

The villagers were insane because of the contaminated water they'd been drinking. The king and his chef were sane because they'd been drinking from an uncontaminated well. The villagers immediately noted, with alarm and concern, that their king was not acting like them. There was something different about his behavior, something not right. They considered whether it was possible to oust the king from his throne.

The king and his chef got thirsty on their shopping expedition. They took a moment to stop at the village well for a refreshing drink. Of course that well was contaminated and they immediately went crazy.

Seeing this, the villagers rejoiced and gave thanks that their king was once again "normal."

We are in a similar situation. "Normal" is the industrial agriculture that is destroying the fertile topsoil we need in order to grow crops for future generations; driving small farmers out of business; contributing to global warming; making it possible for people to be simultaneously obese and undernourished; and creating the conditions for food-borne illnesses. That is crazy . . . unless you've been drinking from that well, and we all have up until recently. We can cut ourselves some slack for not knowing better when we set out on this pesticide-coated, petroleum fuel–driven path. But now we do know better.

Food choices can seem overwhelmingly complicated today. If your prewashed organic salad greens came from 3,000 fuel-burning miles away and are packaged in a kind of plastic that most cities still don't recycle, is that a wise choice (more about local versus organic below)? If your protein bar includes ingredients from several different states and a wrapper that won't biodegrade in the landfill for at least 500 years, is that a healthy choice? And if you're feeding your diabetic family on sodas and burgers because you think you can't afford anything else, even though that food is making your family sick, is that a sane world?

The True Cost of Food

Americans are addicted to cheap food. We are used to seeing low sticker prices that reflect government subsidies and economies of scale. We get a jolt when we see food with a price tag closer to what it actually cost to produce that food.

Our current farm subsidy system was originally intended to protect farmers from the ups and downs of good years and bad. It acted as a leveler, ensuring that the farmer could survive even if there were hailstorms in June or a swarm of pests in August. In the 1970s our farm subsidies underwent a drastic change. Instead of protecting farmers from the vagaries of nature, they became a system of rewarding farmers for growing a small handful of crops, notably corn and soy, in vast quantities. Diversified farms growing a variety of crops and raising a few animals all but disappeared. The monocultures that now make up the landscape of the Midwest took over. Animals stopped being raised on anything resembling farms and were packed into something called concentrated animal feeding operations (CAFOs). The new system of food production was very efficient at producing food in quantity, but the hidden costs to the environment and to the health of both workers and customers have turned out to be unsustainable.

That 99-cent burger at the fast-food place cost a lot more than a buck to produce. Crops had to be grown, animals fed and butchered, workers paid at every step along the line from farm laborer to cashier, and that doesn't even factor in the cost of all the fuel burned shipping the food around or the packaging. But the government subsidizes not only many of the ingredients in that burger but also the fuel it took to get them to the drive-through franchise.

Local and organic foods can seem to carry a higher price tag because we don't see the hidden costs of "regular" food. It's not just that small farms don't get those big government subsidies; it costs the farmer more to grow the food.

As farmer Ted Blomgren of Windflower Farm explains, "We're an organic farm. We don't use herbicides, we don't use insecticides, we don't use fungicides, we don't use fertilizers, so sometimes we run into trouble with insects and with diseases, and all the time we run into trouble with weeds. That's the chief reason organic produce is more expensive than conventionally grown produce. Every one of our solutions is more expensive than the conventional farmer's. For the conventional farmer it usually comes in the form of a pesticide in a jug, and generally they're very inexpensive." (See chapter 6, The Cost Factor, for ways to keep your local, organic diet affordable.)

So who is paying the true cost of all that deceptively cheap conventionally grown food? You are, at tax time when you pitch in your part of the more than $25 billion in agricultural subsidies that the United States hands out every year (and that doesn't include the fuel subsidies). Most of those subsidies go to just 10 percent of our farms, the ones making more than $200,000 a year and with a net worth of nearly $2 million each—in other words, not your small family farm.

The handful of companies in control of the majority of our food supply has billions of dollars to spend on misleading advertising, lawyers to make objections go away, some of the most powerful lobbyists in Washington, and in-house scientists who do product testing that confirms whatever it is that headquarters wants to have confirmed. Against such Goliaths as Monsanto (the company that invented plants that are genetically modified to survive the pesticides that same company sells—one-stop shopping for the absentee industrial farm owner), the odds of what I'm having for dinner helping to save the world might seem slim indeed.

The true cost of food is also borne by underpaid workers exposed to major health hazards from the pesticides they are exposed to every day in the field. Unsubsidized farmers in developing nations also suffer the consequences of our agricultural system when we ship our dirt-cheap subsidized commodities to them and insist that they compete with our prices in the name of "free trade." Because of our farm subsidy policies, we're able to offer food at prices against which local farmers can't compete. In Haiti, for example, farmers once produced enough rice to feed themselves and also export some. Thirty years ago they were forced to compete with the low prices of U.S. subsidized rice and almost all the Haitian rice farms went out of business. When food prices soared in 2008, food riots broke out because people could no longer afford the imported rice and there was no longer a locally grown alternative.

Personally I think there should be a tax exemption for locavores. It really burns me that some of my tax dollars go to subsidizing kinds of food that I simply don't eat, including the corn that goes into that high fructose corn syrup and the meat from animals that never see the sun and are raised in such horrifically unhealthy conditions that they require antibiotics just to keep them alive. I don't want to be paying for cheap rice that put Haitian farmers out of business. I am willing to vote with my dollar and pay the higher, true cost of food for my local and organic groceries, but it seems unfair that I am also subsidizing the artificially cheap prices of food I don't eat.

Food Security

For millennia, humans have founded their cities at locations that provided at least three crucial things: access to drinking water; access to trade via ports or overland routes; and fertile, farmable soil. Every one of our major cities was built near rich soil just waiting to be planted (well, maybe not Las Vegas, but that's another story). That's all well and good so long as a city stays surrounded by that farmland. But in the past fifty years, urban sprawl has paved over close to a million acres of our best topsoil.

Finding affordable land with farmable soil is a serious challenge facing the current generation of would-be small farm owners. In these times, it might make more sense to deliberately plan cities and suburbs on land that *isn't* a good candidate for agriculture . . . just so long as there is good farmland nearby.

While people have always traded for imported goods, until recent times they were also very careful to secure their food supply. What would happen now if there were a fuel shortage or a long-lasting blackout? What would be left on the shelves at your local supermarket or Walmart or Costco if they couldn't restock with long-distance food?

Getting your food from local sources protects your region's farmlands and ultimately secures an immediate food supply. According to the Council on the Environment of New York City, 80 percent of greenmarket farmers say that they would have to sell the farm if it weren't for the farmers' markets. The council estimates that 30,000 acres of regional farmlands have been saved thanks to the farmers' markets in New York City alone. And humans are not the only ones to benefit: The natural landscape that is usually a part of a small farm, adjacent to its crop fields, provides habitat for wildlife.

Food Safety

There is another way small local farms protect our food security. Industrial agriculture does things on a scale that makes the crops and animals vulnerable to disease and infestation. Where thousands of animals are crowded together in unsanitary conditions, or when a single crop is grown over hundreds of acres, anything that goes wrong can spread like wildfire: disease, contamination, pests. Traditional farms were diversified with several kinds of animals and many different crops interplanted. If a pest or disease got to one, the others might still be fine, ensuring that the farmer didn't lose the entire harvest.

Mad cow disease? Swine flu? Salmonella-tainted spinach? Not one of the recent food-related health scares came from a small family farm or from humanely raised, pastured animals.

One of the main reasons food-borne illnesses can become epidemic is lack of

CAN A VEGAN BE A LOCAVORE IN NYC?

When I described my 250-mile diet at a local foods event recently, someone in the audience raised her hand and asked, "I'm a vegan. Could I do a 250-mile diet in New York City?" The answer, sadly, is probably not. A local vegetarian diet that includes eggs and dairy is entirely possible, and in fact more than half my own meals are vegetarian. But some of the main protein sources of a vegan diet are simply not yet available from local farms.

There are no locally grown soy products aside from the edamame that occasionally pops up at farmers' markets. There is locally made tofu, but it is made from nonlocal soybeans. There are no commercially available nuts or nut butters or tahini or sesame or sunflower seeds. What is frustrating is that there is no reason someone couldn't be offering those things here, but no one is. For example, wild nut trees including walnuts, hazelnuts, and butternuts grow in our forests. There's no reason we shouldn't be able to get local nuts and nut oils except that they are labor intensive to collect from the wild and no one has experimented with growing them commercially here. Sunflowers grow beautifully in our region, so in theory we could have sunflower seeds and oil, but no one is doing that . . . yet.

Keep in mind that while a vegan diet definitely has a lower environmental impact when compared with a diet that includes factory farmed meat, it may not when compared with a diet that includes local, sustainably raised animal foods. As local foods activist Liz Neves of www.raganella.com recently asked, "Which has a greater environmental impact: packaged and processed vegan cheese that's shipped across the country or locally, humanely, pasture-raised cow's (or sheep's or goat's) milk cheese purchased at the farmers' market?"

What does all of this translate into as far as planning dinner? For myself, I do continue to eat meat and animal products. I appreciate the artisanal cheeses of our region, the chicken that actually tastes like something, and eggs with bright orange yolks that never break unless I want them to. But I don't eat animal products as often as I used to. I've learned that the way these animals are raised can actually be beneficial for both the environment and my health but only if they are raised in uncrowded conditions. Too many people eating meat too often makes those conditions almost impossible for farmers trying to keep up with the demand. My other reason for eating fewer animal foods is that they are the most expensive items on the locavore's shopping list (more about this in chapter 6, The Cost Factor).

---

traceability. The scale of production on many modern farms is so huge that a single infected plant or animal can spread infection to thousands of products. A single fast-food burger, for example, may contain the meat from hundreds of different cows that were all processed together. If one of those cows was diseased—well, you get the picture.

The labels on milk and egg cartons and on the plastic-wrapped meat at the supermarket still show cheery red barns, white picket fences, and a happy cow or chicken grazing outdoors. That is the stereotype of "farm" stored in our cultural collective unconscious, and it is comforting to believe that is still how the animal products we eat are produced.

The reality is so different from those idyllic images that it has to be hidden away from sight and smell. Conventionally raised animals are kept behind walls, so crowded that they are wading in their own feces, diseased, and requiring antibiotics just to survive. If you think seeing an animal butchered might turn your stomach, seeing how most farm animals are raised today is much more horrifying.

When animals are raised in such unnatural conditions, they become an environmental hazard in several ways. The amount of manure they produce is more than the local soil can incorporate, and the methane gases produced are a significant contributor to greenhouse gases. They aren't grazing open pasture; they're eating food, usually grains, that was grown specifically for animal feed, using acres of land that could have been used to grow food for humans. That animal feed is grown with high concentrations of pesticides. The waterways near animal factory farms are heavily polluted with those pesticides, as well as with the antibiotics and growth hormones the livestock are given.

The feed the animals are given in those CAFOs is nothing nature intended them to eat. Cows, for example, are genetically designed to eat grass, not grain. Eating a grain diet for too long will kill them. The current system solves that problem by killing the animals before their corn diet does.

Other solutions to cheap animal feed are even more macabre and have led to gruesome results. Take mad cow disease, for example. According to a British inquiry into bovine spongiform encephalopathy (BSE, the scientific name for mad cow disease), the epidemic was caused by feeding cattle the remains of other cattle in the form of meat and bone meal, as well as feeding infected protein supplements to very young calves. In other words, these proteins were infected, and the infection was spread to live animals by feeding them the infected proteins. Not only does the cannibalistic practice of feeding cows to cows sound counterintuitive and just plain wrong to me, it also goes against biology. Cows are designed to be herbivores. They don't eat meat. Significantly, cows raised on small farms eating the grasses they were meant to eat did not get mad cow disease.

When animals have enough space and outdoor access, they can actually contribute to the health of a farm. Nancy K. Brown of Lewis Waite Farm provides beef and pork for many CSAs. When asked whether she agrees with the United Nations that animal production is one of the "most significant contributors to the most serious environmental problems at every scale from local to global," she replies:

> I believe that animals in their natural habitat, with enough room, are better caretakers than we are sometimes. They are taking from the environment what they need to live and no more. The cows travel in narrow paths to keep the grasses available for eating; in some cases this reduces erosion, as they are not all walking all over everything. Rotationally grazed pastures (and hayfields in winter) do not need the chemical fertilizers or herbicides, which are some of the greater sources of soil and water pollution. The methane problem is largely generated at the large intensive farms and is not a problem when the animals have space, pastures, and enough room. The dung beetles take care of the rest. But the dung beetles cannot likely survive in a manure lagoon. This applies to most all the grazing animals. The trick is to have enough available land for them.

The animal production the United Nations was talking about was feedlot-raised, grain-fed livestock. The solutions are simple: If you eat meat, dairy, or eggs, get them from small local farms and from pastured animals. Pastured and grass-fed beef, for example, requires half the energy input of feedlot beef and doesn't require the heavy pesticide and fertilizer load that is used to grow corn and soy for conventional animal feed. Be sure that your meat comes from pastured, grass-fed animals.

In addition to getting my animal products from sustainably raised animals only, I also eat less meat than I used to. The reason for this is the space issue Nancy mentions above. In order for livestock to be beneficial rather than detrimental to the environment, there can't be too many of them on any one particular acre of land. If we keep eating meat with the frequency and quantity most people are used to, then we as consumers are demanding overcrowded animal farm conditions in order to support our carnivorous ways.

Saving the World by Eating Great Food

The good news is that all you have to do to stop participating in the litany of horrors that is industrial agriculture is to eat the most delicious, healthiest food on the planet; the food that was harvested in its prime right before you brought it home and that was raised humanely and safely. And there's more good news:

There Has Never Been an Easier Time to Eat Local

A decade ago it would have been nearly impossible for me to do my 250-mile diet unless I moved out of the city and became a farmer (or at least a gardener with a really huge garden and nothing else on my schedule). We have choices now that simply weren't available until recently, and we have them because enough individuals are questioning where their food comes from and demanding sustainable, healthy options. As Pollan writes in *In Defense of Food*:

> *. . . before the resurgence of farmers' markets, the rise of the organic movement, and the renaissance of local agriculture now under way across the country, stepping outside the conventional food system simply was not a realistic option for most people. Now it is. We are entering a postindustrial era of food; for the first time in a generation it is possible to leave behind the Western diet without having also to leave behind civilization. And the more eaters who vote with their forks for a different kind of food, the more commonplace and accessible such food will become.*

Today it may actually be easier to be a locavore in the city than in small towns. In New York City, for example, there are dozens of farmers' markets, CSAs, urban farms, and even online ordering options for local food.

Market Choices: Local versus Organic

The term "organic" indicates whether toxic pesticides were used to grow a particular food, but it doesn't factor in the amount of fuel burned to get that food to us. In the early days of the organic food movement, before there even was a local food movement, the two big considerations were personal and environmental health. "Environmental health" did not yet include an awareness of climate change. Industrial agriculture dumps tons of pesticides and herbicides into the soil. Some of those chemicals leach into water systems and cling to the food we buy and are toxic both to us and to the ecosystems they pollute. So the organic movement set out to offer an alternative: food grown without synthetic chemicals, basically the way all food was grown up until World War II.

It turns out that what we eat has an even greater impact than those agricultural pioneers realized. Whether food is grown organically is only one of the ways food impacts our health and our planet.

As I mentioned in the last chapter, the average distance food travels from farm to plate in this country is 1,500 miles. But because we live in the Northeast and much of our food comes from the other side of the continent, that distance is often closer to 3,000 miles. Those food miles require enormous amounts of fuel, not just to ship the food to us but also to refrigerate the vehicles and the warehouses the food is stored in. The industrial food industry is one of the top two users of petroleum fuel, right up there with the cars people drive, and accounts for 25 to 33 percent of climate change gases.

Let's say I buy some organic salad mix. Because it is organic it's guaranteed not to be genetically modified and to be grown without harmful chemicals. That's good, right? Not so fast. What about the fuel that was burned to ship it from California to my neighborhood in New York? And what about that plastic packaging it came in that is made out of still more petroleum? Even if I recycle the packaging, it will take yet more fuel to put it through the recycling process.

Choosing between local and organic can be confusing. Here is how I make that decision when I'm buying food.

First choice is always local *and* organic. But let's say I'm standing in the store staring at the produce. I'm lucky because this store actually has signs that let me know where some of the food came from. There are local strawberries, there are organic strawberries, but there are no local organic strawberries. What do I do?

In the case of strawberries, I buy the organic because strawberries are on a short list of foods that have a lot of pesticide residue when they are not organically grown. Pesticide residue is not something you want seasoning your dinner: It has been shown to cause endocrine disruption and fertility problems, birth defects, brain tumors and brain cancer, breast cancer, prostate cancer, and childhood leukemia among other health problems. Yikes! And FYI, pesticides do not rinse off with water.

Let's get back to those strawberries. Since I can't get local and organic today, I buy organic. In this case I am making a decision based on personal health rather than carbon footprint. Although the local berries have a lower carbon footprint because less fuel was burned to get them from farm to store, strawberries are on that list of produce that can carry high pesticide residues. If I'd been choosing a fruit that wasn't on the dirty dozen list, I would have chosen local over organic because there wouldn't have been the personal health concern. In that case I'd opt for the lower carbon footprint of local food.

Here is that list of foods worth buying organic over local, if you have to choose, because they carry more pesticides than other produce.

The Dirty Dozen

These fruits and vegetables have the most pesticide residue when conventionally grown.

1. Peaches	7. Cherries
2. Apples	8. Kale
3. Bell peppers	9. Lettuce
4. Celery	10. Grapes (imported)
5. Nectarines	11. Carrots
6. Strawberries	12. Pears

To sum up: Whenever possible, I choose produce that is local and organic. Anytime I have to choose between local and organic, if it's on that dirty dozen list, I choose organic because of the health risks. If it's not on that list, I choose local because of the lower carbon footprint.

If You Do Just One Thing...

Learn a Few of These Local Foods Talking Points and Spread the Word

When I talk to people about the importance of eating locally grown and raised food, I try not to get preachy about it. There's nothing like a holier-than-thou pontificator to turn people off a subject. But at the same time, I am aware when the subject comes up that I am educating people. I try to share both my passion for the importance of our food choices and my delight in local eating. It helps to have these facts handy. Learn them and you too will be able to explain your enthusiasm for eating local in a way that is clear and appealing.

1. **Save the world.** A regional diet consumes seventeen times less oil and gas than a typical diet of nonlocal food.

2. **Taste.** Instead of food bred for shipping durability and long shelf life, local ingredients are raised for flavor and harvested at their peak.

3. **Health.** The shorter the time from harvest to eating, the fewer nutrients are lost. Food that has traveled thousands of miles and sat on a store shelf for days has less nutritional value than its recently picked local counterpart.

4. **Biodiversity.** Ninety-six percent of commercial vegetable varieties have gone extinct in the past one hundred years. Small farms and home gardeners preserve genetic diversity by growing heirloom varieties

that aren't suitable for conventional agriculture. Each variety has unique traits that can be useful in plant breeding, such as disease resistance. And each variety has unique flavors that will be lost to us forever unless those varieties are kept viable.

5. **Food safety.** One hundred percent of the food-borne illness scares of the past few decades came from industrial agriculture. By avoiding the products of industrial agriculture, you can be much surer of the safety of your food.

6. **Support the local economy.** Money spent on local foods is almost twice as likely to be reinvested within the local economy as money spent at a chain supermarket.

7. **Community.** Conversations are ten times more likely to happen at farmers' markets than at the supermarket. If you join a CSA, you will not only get to meet other environmentally aware food enthusiasts but you'll also get to meet your farmer.

8. **Preserve fertile farmland and wildlife habitats.** Small organic farms restore and revitalize their soils rather than deplete them as conventional farming does. They also protect the green spaces adjacent to the farmed fields, providing habitat for wildlife.

9. **Connect to nature and the seasons.** Eating locally connects you to the natural progression of the agricultural calendar in your area, which puts you in touch with the seasons. Trips to the farmers' markets and time spent gardening or foraging get even city-dwellers to spend more time outdoors.

10. **Support your local farmers.** Farmers who sell directly to customers through farmers' markets and CSAs cut out the middlemen and get retail value for their food, which enables them to stay on their land.

3
Sourcing Local Food

"Where do you get your food?" is a question I hear frequently. Most of the people who ask about my local food sources have been to a farmers' market a few times, and many of them have heard of CSAs. But often they don't realize how much local food is available, not only from farmers' markets and CSAs but through other new sources as well.

When I first moved to New York City, it would have been almost impossible to live here as a locavore. The greenmarkets had just gotten started and were small and infrequent. There weren't any CSAs. There weren't any stores like Whole Foods or Trader Joe's that make an attempt to label locally grown food as such. Even the rather sorry-looking corner in the health food store that had a few organic potatoes and gnarly apples didn't post any clue as to where they came from. There wasn't much variety in the supermarket produce aisles, local or not. I'm pretty sure that if I'd asked the produce manager at the Red Apple supermarket next door to me on Columbus Avenue which farms he got his produce from he would have looked at me as if I were from Mars. At best, he would have been able to tell me who the distributors were, which has little to do with where the food is actually grown.

Fortunately, things have changed for the better. Locally grown food has become a trend, much as organic food did a decade ago, with the result that it is increasingly easy to find.

Farmers' Markets

When people think "farmers' market," they usually just think of fruits and vegetables. They may not know that you can also get local honey, flour, wine, every kind of dairy product, maple syrup, meat, seafood, eggs, and other foods there.

I love going to the farmers' markets so much that sometimes I visit them even though I've just picked up my CSA share and don't really need any more food for the week. The atmosphere is like a festival or state fair, especially when the weather is nice (though I do go on the dreariest winter days, too, just to support the stalwart farmers that are out there year-round). It's a pleasure to wander through the colorful stalls and see what new vegetable has appeared for the first time that season, to introduce myself to the farmers and their assistants and chat with them about how they grow their food, to see fellow New Yorkers outside smiling as they stock up on seasonal bounty. I'm not the only one who finds the farmers' markets magical. According to Brian Halweil, senior researcher at Worldwatch Institute, people engage in ten times more conversations at farmers' markets than during supermarket visits.

Most, but not all, of the farmers' markets in New York City are part of the Council on the Environment NYC's greenmarket program. The greenmarkets began in 1976 with just twelve farmers in an empty lot. There are now greenmarkets every day of the week in all five boroughs. At the biggest one, the Union Square market, there can be as many as eighty farms selling their wares on a single market day. And the food available puts the supermarkets to shame. There are literally thousands of varieties of fruits and vegetables, including over one hundred varieties each of apples and tomatoes. This is not just great news for cooks and eaters. According to the UN Food and Agriculture Organization, 75 percent of agricultural genetic diversity was lost in the twentieth century. The heirloom produce and heritage livestock varieties offered by local farms are restoring and preserving our agricultural heritage.

Some farmers' markets are open year-round, others just for the peak harvest season (typically June through November). All the food at the farmers' markets is local, but not all of it is organic. Farms that have an organic certification tend to display that fact on their signs. Other farms may use organic methods but haven't gotten the pricey certification. They may post signs that read PESTICIDE FREE. Still others use farming methods such as integrated pest management (IPM), which means they use organic methods most but not all the time. If you don't see any signs posted that make it clear how a particular farm's crops are grown, it is worthwhile to ask.

The farmers' markets are keeping small farms in business. The absence of middlemen means that rather than the 9 cents on the dollar that farmers get for supermarket food, at farmers' markets the farmer receives an average of 80 cents for each dollar.

The benefits of the markets don't stop there: Neighborhood businesses report a substantial increase in customers when there is a farmers' market nearby.

(See Useful Resources for Web sites that will help you find a farmers' market in your area.)

Grow or Forage Your Own

The ultimate local food is the food you grow or forage yourself. And you don't need to have a garden in order to do so: Community gardens and even windowsills can provide you with fresh herbs and vegetables from spring through fall. (See chapter 5, The Zero Miles Diet, for more on growing your own food.) Wild edible plants grow all around us, even in the city, and learning to correctly identify them can provide you with gourmet ingredients for free. (For more on foraging for wild edible plants and mushrooms, see chapter 10, Feasting for Free.)

In recent years there has been a jump in attendance at the classes I teach on vegetable and herb gardening, as well as for my foraging tours. Other gardening and foraging teachers tell me they are experiencing the same surge in interest. Instead of viewing home food growing and wild edible plants as fringe topics or amusing hobbies, people seem to get it that this is about being able to take care of oneself and one's family directly.

Community Supported Agriculture

Community supported agriculture (CSA) programs began in Japan thirty years ago when a group of women became concerned that the increase in imported produce was causing a rapid decline in the farming population. They decided to create direct relationships with their local farms and called their project *teikei,* which translates loosely as "seeing the farmer's face on the food." The teikei concept spread to other countries, and in the United States the number of CSAs has grown from just 50 in 1990 to over 1,000 today.

CSAs are farm shares in which you share both the risks and the bounty with the farmer. Members pay an annual membership fee up front, in exchange for which they get a weekly delivery of super-fresh food from a local farm. Although the full year's membership is technically supposed to be paid up front, most CSAs offer flexible payment plan arrangements. At the CSA I belong to in Park Slope, we have the option to pay our annual membership in three installments, and there are discounted memberships for low-income families and individuals paying with electronic benefit transfer (EBT). Other options include sharing a share or getting an every-other-week share for single people who are concerned that a full share might be more food than they could keep up with. (See chapter 11, The Single Locavore, for more about these options.)

What the farmer gets: income for seeds, equipment, and crew before there are crops ready to bring to market. In traditional agriculture, a farmer is likely to take out an expensive loan to get started early in the season and then be devastated if it turns out to be a season with lousy weather or other natural catastrophes. This is where the "share the risk" part comes in. The farmer has your membership money regardless of whether it turns out to be a bumper crop year or a lean one.

What you get: gorgeous produce picked within a day or two of delivery (sometimes the same morning), the chance to meet the person who grows your food, and the chance to provide feedback on what you'd like them to grow more or less of next year. You also get to meet other people who love food and care about how their food choices impact the environment.

Each CSA works out its own price schedule with the farmer and its own rules for memberships. Most CSAs require members to work some volunteer time in addition to paying their membership fee. At the Park Slope CSA, for example, members are required to work five hours total, but that is usually divided into two shifts that can be done at any time during the distribution months. The work includes helping the farmer unload the truck, setting up tents if rain is likely, and organizing the week's share so that it is easy for members to pick up. A work shift at a CSA distribution often turns into a fun exchange of recipes and stories about what-I-did-with-all-that-kale-last-week stories.

Some CSAs, including Added Value in Red Hook, offer free shares in exchange for time spent working on the farm. There are also administrative tasks that need doing, and most CSAs offer free or discounted shares to core members who do more than the few hours of work required of other members. (See chapter 6, The Cost Factor, for more on these options.)

The average cost of a CSA membership is around $20 a week. Most CSAs advertise an approximate weight of food that you can expect to get each week. At my CSA, the numbers are 2¼ pounds of leafy greens and 6¼ pounds of "hard vegetables" (root vegetables, squash, etc.) plus "extras" such as fresh herbs. In my experience, how much food you get depends on the season. The very first pickups in late spring are often so light that new members grumble that membership is too pricey for what they're getting (they have no idea yet of the abundance just around the corner). The first week's share is always very, very green: lettuce, scallions, a cooking green like bok choy. But CSA members coming home from a distribution in late August often resemble pack mules, their bags laden with squashes, plums, tomatoes, potatoes, onions, eggplants, and corn.

In addition to the basic vegetable share, members have the option to sign up for fruit, egg, and/or flower shares. (I used to think the flower shares were a frivolous extra. But then I found myself staring enviously at the gorgeous bouquets of organically grown blooms other members walked away with every week. Eventually I caved and signed up for my own flower share. I think of it as locally grown food for my soul.) You can also get meat, poultry, cheese, honey, maple syrup, and bread. These come as a monthly order separate from the regular shares and are optional each month. You can change your order for these products each month or not order at all. Some CSAs have recently added options to buy dry beans and flour milled from local grains.

Some people have trepidations about joining a CSA because they don't have any say in what fruits and vegetables they will get each week. Basically you get whatever the farmer has that week. At our CSA we've got a swap box in which members dump whatever they don't want and take out whatever they like better. I don't use the swap box much. I actually like the surprise factor of each week's selection. What will I be eating this week? I have no idea until I see what we get from the farmer. The CSA puts out a weekly newsletter with recipe suggestions for what we are likely to get that week, and if I'm ever really stuck for what to do with a particular vegetable, the Internet is a great resource. (See chapter 8, Making Friends with your Kitchen, for more ways to come up with recipes for what you've got rather than starting with a recipe and shopping for the ingredients.)

If you are going to be out of town and can't pick up your share, you can't get your money back for that week. Remember that the farmer planned his crops around the number of members in the CSA and a guaranteed income from their annual fees. You can, however, have a friend pick up your share while you're away and either enjoy it themselves or save some for when you get back.

Several CSAs have added winter shares to their distribution schedule. These are usually monthly deliveries unlike the weekly distributions of the growing season. In my winter share I usually get apples and cider, eggs, storage vegetables such as beets and potatoes and winter squash, some greens from Farmer Ted's unheated greenhouses, and sometimes dry beans and garlic.

37

CSAs also offer farm trips a couple of times a year. Members get the chance to see where and how their food is grown. Some farms welcome visitors anytime (with a little polite advance notice). Nancy K. Brown of Lewis Waite Farm says that she thinks of her farm's CSA members as "neighbors of a sort, and customers too. If the distance were not so great to our farm for our CSA members in New York City, those members could come and visit more often; we could host picnics and events and really get to know more members as friends. Many of the people who have come to visit, come again and keep in touch."

As at the farmers' markets, not all food from CSA farmers is organically grown, but many of the farms that are not certified organic do in fact practice organic farming methods. Every CSA farmer I've talked to is concerned about taking care of the soil on his or her farm, the plants growing in it and animals living on it, in ways that are healthy and sustainable, whether they are certified organic or not.

To find a CSA near you, contact Local Harvest or Just Food (see Useful Resources). Many CSAs are wait-listed because the demand has exploded in the past few years. I've had a chance to experience the rise in CSA popularity first-hand. When I joined the Park Slope CSA in 2003, we had to do a major recruitment campaign every spring to ensure that there would be enough members to make it worth our farmer's time. We spent a lot of time putting up fliers and explaining what a CSA was, because most people had never heard of one. During the months of distribution, people would stop by thinking we were a farmers' market and wanting to buy a single peach or head of lettuce. We would explain that they couldn't do that because it was a CSA membership, what that was, and that there were still a few memberships available that we'd be happy to prorate them for. They would hear us out but apparently not comprehend because inevitably at the end of our pitch they'd say something like, "But how much for the peach?" Nowadays a simple "Sorry, we're a CSA" gets an understanding nod usually followed by "How do I get on the list for next year?" Starting in 2007 our CSA's membership sold out each year without us having to do any recruitment, and in 2009 we expanded from a once a week distribution to twice a week.

If the CSA closest to you is sold out for the season, or you can't find one nearby, Just Food can help you start one. They'll work with you on the specifics of getting your CSA up and running, including connecting you with a farmer, helping you find a distribution location, and recruiting members.

Community Garden Markets and Urban Farms

City farm markets are community gardens that double as farmers' markets, selling food grown right on the spot. These gardens and markets are thriving in inner-city neighborhoods where fresh food has been hard to find in the past. (It's not unusual in underserved neighborhoods to find half a dozen fast-food chains and not a single supermarket.)

The Queens County Farm Museum is a greenmarket farm located within city limits. In addition to manning a stall at the Union Square market, they sell eggs and produce at the farm and in 2009 began selling wines made from grapes grown on the farm.

BK Farmyards offers a unique service in which they will grow vegetables for you if you have garden space. The way it works is that you set up a consultation with them, approve the start-up cost estimate, and then let them tend your edible garden. They harvest your food when it is at its peak and bring it to you.

Schoolyard and Playground Gardens

Thanks to Alice Waters, many schools on the West Coast now have their own food gardens. Finally the idea is taking off here in the East. In Red Hook, Brooklyn, the group Added Value has transformed a dilapidated playground into a farm worked largely by teenagers who not only get to eat what they grow but also make money from it, thanks to the farmers' market they've started. The playground was completely paved over, but that didn't daunt them. They made raised garden beds in which to grow their vegetables right on top of the asphalt.

The benefits of schoolyard gardens go beyond providing better nutrition for the students. Teachers tie their class work to the students' gardening activities, teaching botany and other topics related to plants and the natural world. The gardens give the students a direct connection to the subjects they are learning in the classroom. And if you think kids don't like to eat their vegetables, think again. Many times I've seen a kid eating a carrot or a bell pepper with pride and appreciation. It's hard to turn up your nose at something that you planted from seed, weeded, watered, staked, and tended for months before it showed up on your plate.

One of the challenges that initially faced NYC schools trying to get food from

39

the school garden onto the students' plates was lack of real kitchen facilities and trained staff. This was because few schools had real kitchens during the past three decades. They had microwaves and chafing dishes, because that was all that was needed to heat the frozen, packaged food they served up. The servers were not trained in cooking skills beyond setting the timer on the microwave ovens, so even if you brought them some terrific parsnips that the students grew, they weren't going to be able to do anything with them. Fortunately that has changed in recent years, thanks to the efforts of organizations such as Food Systems Network NYC.

Food Co-ops

Most food co-ops make an effort to offer some local produce when they can get it. The co-op I belong to, the Park Slope Food Co-op, has long-standing relationships with several nearby farms. During the growing season, I can count on getting local and usually organic fruits and vegetables, which is very convenient on days when I can't get to a farmers' market. The co-op's supply of locally sourced food kicks in about six weeks before my CSA share starts, and long before the garden is producing much. Prices at co-ops are lower than at the supermarket or farmers' markets. (For more about co-ops see chapter 6, The Cost Factor.)

Wine Merchants

Some wine stores carry a decent selection of local wines, others not a single one. There is one popular wine store on the Upper East Side where the manager looked at me like I was crazy when I asked if they had any New York wines. This is surprising considering that our wineries are getting increasing respect and attention from the wine industry. Check out the Locavore's Guide to NYC online for my latest update on which wine merchants have a good local selection. You can also get wine and cider at the greenmarkets, though I have to admit that the quality ranges from excellent to barely drinkable.

By the way, I haven't found any local beers yet. There are beers that are brewed locally, but they are made with nonlocal grains and malts.

Specialty Stores and Cheese Mongers

Many gourmet specialty shops and most good cheese mongers offer some local products. They are admittedly inconsistent though: I was delighted to find one of my favorite local cheeses (a Barat by Sprout Creek Farm) at Eli's on the Upper East Side, only to find out a month later that they weren't carrying it anymore. That has happened to me a few times at different stores.

On the positive side, the cheese monger at Bierkraft in Brooklyn consistently stocks a great selection of local cheeses. Keep an eye out for local selections at the specialty shops, and most important keep asking for local. They're starting to catch on.

Local Food Online

This is not my first choice because products ordered online are often shipped by air, which uses more energy than travel by land. Nonetheless, in many cases a local product shipped by air still uses less energy to get to you than its cousin shipped from across the country or overseas. In the Locavore's Guide to NYC, I include links to farms that offer online ordering.

If you are a Fresh Direct fan, you are in luck because they have a local foods section that includes vegetables and herbs from Long Island, as well as local dairy, poultry, seafood, and wines. If you are going for a pure local foods diet, be aware that not everything marketed as local is made with 100 percent local ingredients. For example, Fresh Direct includes potato chips made in North Fork, Long Island, in their local foods section. But although they are made locally using locally grown potatoes, the ingredients list also includes nonlocal oil and salt.

Supermarkets

Most supermarkets still don't post which items are local, or do so only occasionally. But hey, it took them awhile to get on the organic bandwagon too. A little sleuthing will often reveal local food tucked in with the nonlocal stuff. For example, my neighborhood supermarket recently started carrying Natural-by-Nature and Ronnybrook, both excellent brands of local, organic milk. You can probably find maple syrup from New England on the shelves and a few cheeses from the region. If we keep asking for local foods, the supermarkets will come around just as they did with organic produce a few years ago.

U-Pick Farms

If you have a car, there are lots of nearby farms that let you pick your own fruit. Often the prices are cheaper than they would be otherwise because they aren't paying workers to do the picking. Plus, it's a nice way to spend a day or half a day out of the city. The Pick Your Own Web site can help you locate nearby U-pick farms (see Useful Resources).

That's Where to Get It; Here's What You Can Get

The gorgeous procession of fruits and vegetables coming into season includes everything that can be grown in our climate. What that doesn't include is citrus, avocados, and tropical fruits like mangoes and bananas. Because local food is by definition seasonal, you won't find everything all the time. In addition to kitchen staples like potatoes, carrots, leafy greens, and tomatoes, I look forward to the treats that show up for only a week or two, including spring's wild fiddlehead ferns and green garlic, summer's sour cherries (amazing pickled), and the shell beans of late summer and early fall. (To find out which produce varieties you can get when, see chapter 4, Eating with the Seasons.)

Increasingly, our local food includes more than fruits and vegetables. Our local cow, sheep, and goat farmers are pairing with artisanal cheese makers to make everything from blue cheese to ricotta to feta, as well as excellent butter, yogurt, buttermilk, ice cream, and even half-and-half for your (nonlocal) coffee.

As I mentioned above, there are some excellent local wineries. It's amazing how many people still don't know that. I recently invited a colleague over for a local foods dinner and mentioned that we'd be enjoying some local wine along with the meal. He e-mailed back, "Local wine? I love that idea! How does one do that?" I think he imagined I was brewing it up myself, which I've done from time to time, but in this case all I had to do was stop in to Red, White, and Bubbly half a block up the street from me.

The two main grape growing and winemaking areas are the Finger Lakes region in upstate New York and the North Fork end of Long Island, although there are also an increasing number of winemakers in the Hudson Valley. And while we're sipping something alcoholic, the local hard ciders are also excellent.

Speaking of cider, our local (sweet, aka nonalcoholic) apple cider is a treat I look forward to every fall. Several farms are making fruit juice blends that include raspberries, grapes, cherries, and other fruit. They are delicious, and you won't miss that glass of nonlocal orange juice you used to have with breakfast.

Although all the bread and baked goods you'll find at the farmers' markets are locally made, not all of them are made with local grains. But there are a few farms growing and milling grains. I always thought of wheat as something that had to be grown on the big, flat plains of the Midwest until I discovered Wild Hive Farm's bread flour. Since then I've baked with their all-purpose, pastry, whole wheat (in various forms from coarse to fine), rye, spelt, and triticale flours. I've also found other grain farmers and millers, including Ted Blew of Oak Grove Plantation. In addition to wheat flour, other locally grown grains include oats (and really great oatmeal!), rye, kasha, and popcorn.

Mushrooms are available year-round, not just at the farmers' markets. Most of the commercial mushrooms in our area are grown in Pennsylvania, so this is one local food you can easily find at any supermarket. Outside the supermarket, you can find some of the more interesting (though pricier) gourmet varieties, including oyster mushrooms, maitake, and porcini. And if you learn safe field identification, all of those mushrooms and more grow wild all over the city. (Check out the Useful Resources appendix for where to find classes and tours with wild edible plant and mushroom experts.)

Speaking of wild edibles, there are so many choice ones just waiting to be harvested that I've devoted a chapter to the subject (chapter 10, Feasting for Free).

When I did my 250-mile diet year, it took me four months to track down dry beans. Until then, I was enjoying the shell beans at the markets. (Shell beans come in the pod and need to be shelled, hence the name. They are the size of their dry counterparts but have a more tender texture and a much shorter cooking time. Left on the vine, they eventually become dry beans.) The source I eventually found was Cayuga Pure Organics. They are still offering several varieties and have started selling them through the city's CSAs. I'm happy to report that in the time since my 250, more farms are offering dry beans at farmers' markets, online, and through CSAs. It's much easier to find them now, which is especially good news for vegetarian locavores trying to meet their protein quota.

If you're not a vegetarian, you have a wealth of locally raised meats, seafood, and poultry to choose from. We can get beef, lamb, and even goat. Pork products include everything from chops to bacon to sausage. Chicken, duck, quail, pheasant, and turkey are all raised locally and are available in every form from whole to ground. The local eggs are incomparable—you won't be able to go back to eating any other kind once you've seen their bright orange, perky yolks and seen what they do for every recipe you use them in. (Baked goods really do rise higher with fresh eggs, by the way. I used to read previous generations' cookbooks and think it was quaint when they specified using "the freshest eggs," but they were right.)

The longest line at the farmers' market is almost always for the seafood stall. If you come even an hour late, you'll find that they've sold out of much of their catch. Numerous kinds of wild-caught fish and shellfish are available, as well as calamari. All of them are sustainably fished within 100 miles of the city and are unbelievably fresh and delicious. You can also get a freebie from the fishmongers (see chapter 6, The Cost Factor).

Fruits and fruit juices, vegetables, dairy of every kind, eggs, beans, grains, wine, cider, mushrooms, meat, poultry, seafood . . . have I left anything out? Yes, I want something to season my cooking with. Fresh herbs and chili peppers (fresh or, when out of season, dried) are available year-round. So is garlic in one form or another. The garlic progression starts with the wild field garlic of early spring. Then come the garlic scapes and green garlic, followed by the main garlic harvest of July and then the stored garlic bulbs that are available well into winter (these last are the closest to what we are used to getting at the store). There is no black pepper (a tropical vine mostly grown on the Malabar Coast) or commercial ginger (an Indonesian grass), but there is wild peppergrass and a wild ginger that is a native northeastern woodland ground cover available to foragers (see chapter 10, Feasting for Free).

4
Eating with the Seasons

I turn on *New York One* and hear a celebrity chef touting his new restaurant. "And of course it will feature local, seasonal ingredients because that's the best food there is," he says. "Local" and "seasonal" are said in the same breath so often nowadays that they've become buzzwords, and for good reason.

If I buy an eggplant in March, I can be sure it came from South America or Florida or somewhere with drastically different weather patterns from where I live because eggplant is only in season in summer and early fall here. There is no such thing as a nonseasonal local food, with the exception of hothouse vegetables. You can buy hothouse tomatoes and peppers at the greenmarkets even in winter, but hothouses burn petroleum fuel to keep the plants warm, and since one of my primary reasons for eating local foods is to reduce the amount of fuel used to produce my food, it doesn't make sense to me to eat hothouse food. Plus, those tomatoes don't taste nearly as good as the field-grown, in-season ones. So although hothouse vegetables are technically local, I'm not a fan.

My friend Bill lives near the Union Square Greenmarket. He called one wintry day to let me know that my enthusiastic promotion of local foods had inspired him to go shopping at the market. "But there wasn't much there," he told me. "Well, it is January," I said. "What did you expect?" "Um, everything I saw when I was there last summer?" he replied, laughing at himself as he said it.

Bill's not alone in his ignorance of seasonal eating. Even my father, who grew up on a farm in Indiana but has lived most of his life in San Francisco, admits that he isn't sure what is in season when. In a world in which almost every food is available every day of the year, it is no wonder so many people have lost the knowledge of when things naturally grow and ripen in their climate. I used to be the same. If eggplant was on my shopping list or I saw it at the supermarket and had a sudden craving for eggplant Parmesan, I'd buy it without a second thought.

We've divorced ourselves from the natural progression of the seasons in so many ways, not only with food. We spend much of our lives indoors where the air is heated in winter and chilled by air conditioners in summer. Sometimes it seems that our closest connection to the natural world is checking the Weather Channel to see if we need to take an umbrella with us today.

Still, all it takes is one glorious spring day and every restaurant in my neighborhood brings out its outdoor tables. "Have you gone to see the cherry blossoms at the Brooklyn Botanic Garden yet?" I hear one woman ask her friend as they step into a salon for the first flip-flop pedicure of the season. We do still respond to the seasons beyond basic choices, such as whether we need a coat today. The same inexplicable joy we get from April's cherry blossoms or October's spectacular fall foliage is there for seasonal eaters in every new ingredient as it reaches its peak and then steps aside to make way for the next coming-into-perfection food.

Learning, or relearning, what is in season when can be tricky. Even at the Park Slope Food Co-op, which prides itself on supporting local farms whenever possible, there is not much change when you might expect it. In August the produce aisle is stacked on either side with a rainbow of tempting fruits and vegetables. There are orange and yellow bell peppers, green zucchinis, red tomatoes, purple kale, blueberries, and blackberries. The co-op makes an effort to clearly label where each item originated. Since it is August, today the peppers and kale are from the nearby Hudson Valley, the zucchini from a little farther north in the Finger Lakes region, the fruit from Pennsylvania.

Fast-forward to January. The aisles look almost exactly as they did last summer: The colorful selection is still there to tempt me. The difference is that now the produce is labeled as coming from Mexico, Peru, Italy, Florida, and the vague and vast "USA." As in the typical supermarket, the co-op stocks almost everything all the time because that is what consumers expect.

The prices are higher, though, and those imported tomatoes lack the luscious color and juiciness of their locally grown, in-season counterparts. They are also less nutritious, since the vitamin content of fruits and vegetables starts to diminish the moment they are picked. These were picked many days ago and then on the road for a while before arriving on the shelves. So yes, I could have that out-of-season tomato any time I want to, but why would I want to? If I wait for the real thing, I will be rewarded with something many times more delicious, less expensive, and packed with more nutrition. Meanwhile, there are the jars of home-canned and dried tomatoes that I prepared when tomatoes were at their peak during late summer: Those are my winter tomatoes.

Even people who are up on all the reasons seasonal food makes sense to cooks and environmentalists alike sometimes fall off the wagon. On a recent winter's visit, my mom, who surely has heard me pontificate about the virtues of local, seasonal food often enough to know better, grabbed a late December tomato off the stacked pyramid of pallid produce at the supermarket. "I know, I know," she said, "but I just want it, I just do." So she bought it and sliced it up for that night's salad. At dinner she said, "This tomato isn't very good." I raised an "I-told-you-so" eyebrow and she giggled guiltily.

One of the unexpected benefits of my 250-mile diet year was that it cured me utterly of my addiction to the "anything anytime" eating pattern that has become our norm. Since by my own rules I couldn't touch the out-of-season stuff, I didn't. As a result, each new ingredient in the seasonal parade was, well, new to me. When the first asparagus arrived from South Jersey in April, and I hadn't had any for a year, walking wasn't fast enough to contain my excitement. I skipped over to that farmer's stall, just a few skips, but they were enough to earn me a scolding "grown-ups aren't supposed to do that" expression from one angel-faced toddler. For the remainder of asparagus season, I ate it at least every other day: steamed, grilled, roasted, raw. Yeah, I was actually getting a little tired of it by the time it was going out of season. But right then the strawberries hit their peak. (Skip, don't walk, to that first perfect strawberry of late spring!)

Each seasonal ingredient was so spectacularly good that the alternative wasn't even tempting. I'd look at an out-of-season strawberry and it didn't look right, it didn't smell right, and I knew it wouldn't taste right. Let me get this straight, I'd think, I could pay a premium price in both dollars and environmental cost and what I'd get wouldn't even be very good? It just stopped making any sense. Nonetheless, I know that before The 250, I was still sometimes lured by my expectation of something wonderful, a stored memory of what a strawberry could be. I'd spring for that out-of-season fruit, only to be underwhelmed and unfulfilled when I ate it.

Convincing people that if they start eating seasonally, they not only won't feel deprived but will be eating much more delicious food and won't get bored because what's available will change every month is like telling a junkie that she really will feel better when she kicks the habit. The facts and the evidence all line up, but the junkie's gut instinct is still to reach for the bad stuff. One of the many things I'm grateful for about having done a 250-mile diet for a whole year is that it cured me of the craving for out-of-season foods without my ever realizing that I was in recovery from the way most people eat.

Liz Neves, a Brooklyn sustainable living consultant and blogger at www.rag anella.com, says, "One thing I appreciate about eating locally is what goes hand in hand with it, eating seasonally. I know it's beneficial to eat certain foods at specific times of the year. In winter our bodies slow down, and so does our digestive system. Apples, generally associated with autumn and winter, aid digestion. It's also exciting to anticipate the season's harvest. We eat asparagus in spring, fresh strawberries in summer, squash in fall and winter. That's the way it's supposed to be!"

South to North

As I sat at my computer calculating farm-to-plate miles for the Locavore's Guide to NYC (e.g., Keenan Farm, 108 miles; Phillips Farm, 59 miles), I became aware of small differences in climate that affected what would be available to me each week. The first asparagus comes in from South Jersey, which warms up a couple of weeks sooner than Long Island and almost a month ahead of the Hudson Valley and Finger Lakes regions. When those South Jersey farms shift from asparagus to the early-summer crops, the asparagus crops of farms farther north are just starting to kick in. This is good news for NYC–area locavores, because it means that we have an extended season for each fruit and vegetable, starting with South Jersey's offerings and ending with those same crops coming in a few weeks later from farther north.

It seems obvious that southern regions have warmer weather and longer growing seasons than northern. But that scientific fact is part of the environmental impact of nonlocal foods today. When Europeans came to North America, they were able to bring with them many of their food staples because plants travel well moving from east to west or vice versa. Apples, chard, leeks, and other European crops grew just as well in our climate as they had in Europe.

Plants don't travel nearly as well moving from south to north because of the difference in winter temperatures. There never will be a mango tree that can survive a Brooklyn winter. When someone from South America re-creates his or her food culture here, it includes ingredients that have to be imported because they will not grow here. That is a very different ecological situation from that of Italian immigrants, who were able to plant cold-hardy European grapes in their American backyards.

When I posted about this on my blog, I got slammed by a few indignant readers. They thought I was suggesting that Europeans somehow had more of a right to import their food culture to North America than other people. That was absolutely not my point. I was pointing out that the scientific reality is that plants can adapt well to new terrain if they've been moved east–west but not if they've been moved south–north. It's a question of whether they can survive the winters of a particular region. But the ruckus my post caused points out an important issue: Food is not just about crops and nutritional value. It is also about people, their memories, their recipes, and their culture. Tricky.

A Reverse Approach to Recipes

After decades of having everything available all the time at the supermarket, people have gotten used to a recipe-first approach to cooking. With the recipe-first approach, you decide what you're having for dinner, make a shopping list based on your chosen menu, and then hit the store to get the ingredients you need. Never mind that those ingredients may not look very good that day.

This is the opposite of how people have traditionally cooked. Actually, it's the opposite of how most other cultures still cook. In most places in the world, it is still the norm to see what looks good that day and then plan a meal around it. That's the reason refrigerators and freezers in other countries are often half the size of ours in the United States. Other cultures are still not used to the idea of doing a massive shopping and stocking up for the whole week or month. Instead they shop every day or every few days, looking for what is freshest and best.

Most of the fruits and vegetables that sit on the shelves of the average supermarket come from a limited number of varieties. There are hundreds of kinds of tomatoes, for example, but at the store you'll probably only see three: the big round slicer, a pear-shaped Roma variety, and cherry tomatoes. These varieties were bred for durability during shipping and long shelf life. The trade-off is what they were not bred for: taste and nutrition. All those hundreds of other varieties of tomatoes are delicious, but they bruise easily when ripe and don't last long on the store shelves.

Speaking of tomatoes, don't be fooled by the ones that are labeled "vine ripened" and come still attached to a section of vine. Vine ripened is not the same as sun ripened in the field. It is absolutely possible to cut off a length of vine with green tomatoes clinging to it and wait for those tomatoes to turn red, "ripening" on the vine. But although they turn red, they don't continue to build up the sugars and nutrients that they would have if they were still connected to the parent plant and growing in the sun.

Eating seasonally means reversing the process of plan meal/make shopping list/go shopping. Instead it becomes go shopping (or step out into the garden), see what is good, then decide what's for dinner. Instead of planning a meal and then praying that I can find what I need from a local source, I look at what I've got and let it inspire me.

What's in Season

Here is a list of what food is in season when in the Northeast. Keep in mind that crops are affected by weather variations such as cooler-than-usual springs, so actual harvest dates can be a week to ten days earlier or later. A few farms use unheated greenhouses so that they can offer leafy greens for more months than are listed here.

Vegetables

Vegetables	Harvest	Availability
Asparagus	Late April–mid-June	Late April–mid-June
Beans, lima	Late August–October	Late August–October
Beans, snap	July–October	July–October
Beets	June–November	June–February
Beet greens	May–September	May–September
Broccoli	June–November	June–November
Brussels sprouts	September–November	September–November
Cabbage	June–early December	June–March

Carrots	July–November	July–early April
Cauliflower	Late August–November	Late August–November
Celery	August–early November	August–early November
Collard greens	July–December	July–December
Corn	Late July–October	Late July–October
Cucumber	July–early November	July–early November
Eggplant	Late July–early November	Late July–early November
Garlic, bulbs	June–July	June–November
Garlic, green	Late April–May	Late April–May
Garlic, scapes	May–early June	May–early June
Kale	August–November	August–November
Leeks	August–October	August–December
Lettuce	May–early October	May–early October
Mustard greens	May–early October	May–early October
Onions	July–November	year-round
Parsnips	October–December / April–May	October–May
Peas	Late May–July / September–October	Late May–July / September–October
Peppers	July–early October	July–early October
Potatoes	July–October	year-round
Pumpkins	September–October	September–November
Radishes	Late April–September	Late April–September
Ramps	Late April–May	Late April–May
Rhubarb	May–July	May–July
Spinach	April–December	April–December
Squash, summer	June–early October	June–early October
Squash, winter	August–early November	August–February
Swiss chard	June–October	June–October
Tomatoes	July–October	July–October
Turnips	August–November	August–February / young turnips in June
Turnip greens / Baby turnips	May–August	May–August

Fruits

Fruits	Harvest	Availability
Apples	July–November	year-round
Blackberries	August–September	August–September
Blueberries	mid-July–August	mid-July–August
Cantaloupes	August–September	August–September
Cherries, sweet	Late June–July	Late June–July
Cherries, sour	July	July
Currants	June–August	June–August
Elderberries	August	August
Grapes	Late August–October	Late August–October
Nectarines	July–early September	July–early September
Peaches	July–early September	July–early September
Pears	Late August–October	Late August–February
Plums	July–September	July–early October
Raspberries	early July / late August–October	early July / late August–October
Strawberries	Late May–July	Late May–July
Watermelon	Late August–October	Late August–October

Seasonal Seafood

As a locavore, gardener, and forager, I used to feel a little bit smug about how much more aware I was of what was in season than most people—until I realized that my knowledge only applied to fruits, vegetables, and mushrooms. When it came to other foods, I was utterly ignorant.

I found this out after I fell in love with fluke, a summer fish that is fairly cheap and has a succulent texture and slightly sweet taste. I bought fluke every week from Blue Moon Fishery at the Grand Army Plaza Greenmarket in Brooklyn until one Saturday when they didn't have it. When I asked, "When will you have fluke again?" the fish monger gave me one of those bland, patient looks I've come to know translates as, "Ah, these clueless city folk." "Next summer," she said curtly; "Fluke is a summer fish."

Well, I guess I'd known that seafood had seasons. I knew about salmon runs and that there were months you weren't supposed to eat shellfish (the ones with an "R"? without an "R"?), but I hadn't paid much attention because the fish I'd eaten most of my life was frozen and then thawed imported fish that was available whenever.

Many kinds of fish migrate just like birds do. They move up and down the coast as the ocean water temperature changes. Some of them move through our area in spring or summer as they migrate north and again at the end of summer and in early fall as they move south. The exact timing of these migrations varies slightly from year to year depending on seasonal weather patterns and ocean temperatures. Availability is also influenced by fishery management regulations that are designed to keep fish populations stable. These include size limits to ensure that individual fish live long enough to reproduce before they are caught. Some of the regulations prohibit fishing at times when fish are spawning or reproducing. All of these factors influence what seafood we can get when.

Here's a partial list of what our local fish mongers have, depending on the season:

Spring	Summer	Fall	Winter
Atlantic mackerel	Blue crab	Bay scallops	Cod
Blackfish	Bluefish	Blackfish	Hake
Butterfish	Butterfish	Bluefish	Herring
Cod	Conch	Butterfish	Ocean pout
Flounder	Flounder	Cape shark	Whiting
Hake	Fluke	Conch	
Jonah crab	Jonah crab	Flounder	**Year-round**
Lobster	Lobster	Jonah crab	Clams
Monkfish	Mahimahi	Lobster	Oysters
Ocean pout	Mussels	Monkfish	Squid
Porgy	Sea robin	Porgy	Tilefish
Sea bass	Sea scallops	Sea bass	
Sea scallops	Shark	Shark	
Shad	Striped bass	Skate	
Skate	Swordfish	Striped bass	
Whiting	Tuna	Swordfish	
		Tuna	
		Weakfish	

Seasonal Dairy

My understanding of seasonal food got another upgrade when I found out that the feta cheese from my CSA wasn't available year-round. My favorite feta is made from either sheep or goat milk. Well, it turns out that sheep and goats don't lactate year-round. They take a couple of months off during winter. Since feta is a fresh cheese rather than an aged one, it is made during the lactating months only. So the best time to eat fresh, unaged sheep and goat's milk cheeses is during spring and summer.

Cows have the longest lactating period, so by staggering when the cows in a herd calve (remember, animals produce milk to feed their young, not us), it is possible to make cow cheeses all year.

The seasons affect the taste of an animal's milk because of changes in the animal's food. In spring a pastured animal will be eating young grasses and flowers that give floral and herbal tastes to the milk and to the cheese made from spring milk. In summer the grasses have a higher beta-carotene content, which changes the taste of the milk. In winter the hay and grain the animals eat changes the taste of their milk yet again.

Certain cheeses are considered best made from the milk of specific seasons. For example, blue cheeses are considered best when made with summer milk and then aged three to five months, so the best time to buy blue cheeses is in late fall and early winter.

Aged cheeses are available year-round, but if you pay attention you'll notice differences in taste and texture among a variety of cheese purchased in August, for example, versus that same type of cheese in October.

Knowing What's in Season Means Knowing If It's Local

Once you know what grows near you when, you know instantly if a food had to be imported. Eggplant in New York in March? I don't think so!

My friend Ellen recently went to a well-known restaurant in Manhattan that advertises its support of local, seasonal food. This was in late spring, and one of the dishes on the menu featured hen of the woods, aka maitake mushrooms. Ellen called the waiter over. She apologized and said she wasn't trying to be difficult, but there was no way the restaurant was getting that particular mushroom locally because it only grows here in the fall. The waiter assured her that the chef only used local, seasonal foods on the menu. She insisted, and he went back to the kitchen to check. When he came back he said, "Well, actually it turns out that we're getting those from Japan right now." Knowing what's in season is key to knowing what's local!

IF YOU DO JUST ONE THING . . .

Learn when your favorite foods are in season.

Take a look at the charts above and highlight your absolute favorite fruits and vegetables along with their harvest dates. This is when they are going to be at their most delicious, cheapest, and most nutritious. Enter them into your calendar and plan on stocking up then. Pass on the imported, bland, out-of-season supermarket versions that are available at other times.

55

5

The Zero Miles Diet:
Grow It Yourself

Food that you planted, watched grow, just picked, and are carrying back to your kitchen is the ultimate local food. Nothing else compares when it comes to flavor, nutrition, minimal environmental impact, and the sheer joy of being an active partner with nature on the project of feeding yourself. Its only match is the wild edible plant you just correctly identified and collected.

If you don't have a garden but would like to grow some of your own food, take heart. There is a gardening space available to you even if you live on the twentieth floor of an apartment building. I have grown food on fire escapes, indoor windowsills, rooftops, in community gardens, and finally, blessedly, in my own backyard. It's gotten to where I can't imagine not doing so. You'll find information on locating a garden if you don't have one and on how to grow food in it below.

For those of us who will never dwell on a family farm, the reasons for growing at least a little of our own food are many. It can be a way to keep food costs down, to teach kids firsthand where their food comes from, and to grow special varieties that aren't often available commercially. Hands-on interaction with plants, even in an urban setting, is healing on many levels, so much so that the New York Botanical Garden and other organizations offer courses in horticultural therapy. In so-called "underserved" neighborhoods (i.e., low-income neighborhoods where the only sources of food may be convenience stores and fast-food chains), food gardens provide an affordable way to get much needed fresh food and nourishment to residents. Many community gardens are intensively planted for food, but they also provide public access to green space even for those who aren't members of the garden. Across the street from me, for example, is a small community garden where nongardeners regularly hang out, enjoying the plants and the sunshine while they read a book or check their e-mail on their phones.

The garden I tend now is the backyard of a Park Slope apartment building. I share it with the neighbors who live at the same address in the apartment next to mine. When I first moved here, nobody else had access to the garden. It had been neglected for years and was just a square of mangy, patchy lawn and ivy. I dove in with enthusiasm, creating raised beds and borders mostly filled with edible plants. The sunniest part of the garden is adjacent to the building and concreted over, so I planted herbs and vegetables in containers there.

When the landlord turned the basement into an apartment and unlocked the door that leads from the basement to the garden, it became a shared garden. I was nervous that I would have to give up a lot of the space I was growing food in, but mercifully my new neighbors were fine with letting me do most of the planting. They have added some containers this year to the sunny area, and I'm happy to encourage them to add more. Along with the usual suspects (tomatoes, basil, cucumbers) the garden now boasts an edible shade patch with mayapples, fiddlehead ferns, and wild ginger, as well as elderberries, red currants, strawberries, and an interesting fruit called cherry elaeagnus, aka goumi, or gumi.

When I worked as a professional gardener in New York City, I often got comments along the lines of "Gardens? There can't be many of those here." But there are. There are backyards hidden from the street, green rooftops, lush container gardens on terraces, community and schoolyard gardens, pots of flowers on the front steps of brownstones, windowsills brimming with planted goodness. Every building on my block has a backyard—a treasure invisible from the street.

I started gardening in my thirties on the roof of an apartment building. Not knowing anything much, I tried everything. I learned that 6-foot-tall corn is a bad idea on a roof because the wind knocks the plants flat. I learned that butterflies, dragonflies, birds, and (less exciting) tomato hornworms find their way to a rooftop garden even on a block with no trees or other greenery. Indeed, providing a migratory stopping place for wildlife is one of the lovely benefits of a garden. I learned that the rooftop gardener before me hadn't considered irrigation, and that I had to haul water up the stairs in watering cans multiple trips per day.

Fortunately I was more thrilled by my successes than I was daunted by my failures. I loved that roof garden so much that I started dreaming of having some earth to plant in, an actual in-the-ground garden. Since I couldn't afford a garden apartment in Manhattan, I moved to Brooklyn, where I tended a small, shady backyard for eight years. I also joined a community garden. I was still learning by trial and error. (Why hadn't I ever pestered my grandmother with questions when was a kid? She loved gardening, and could have taught me a lot. Oh, well.)

Eventually I got tired of trial and error and started to take some classes at the New York Botanical Garden to learn how to be a better gardener. From there I went on to become a professional gardener for private and corporate clients in the city. And that brings me to an important point: There is no such thing as being born (or not) with a green thumb.

Getting a Green Thumb

"But I killed all my houseplants! I can't garden. I'm just not good with plants."

I hear these objections frequently when I teach classes at the New York Botanical Garden and the Brooklyn Botanic Garden. People take it personally when the plants under their care don't thrive. I think this comes from an excellent instinct: They feel responsible because plants are living things and feel bad if they die on their watch. There are two things they don't know that would make them feel better.

One, gardening is a cocreative act between humans and nature. Even if you do everything right, nature may decide that this is the year your garden will be attacked by Japanese beetles or an especially bad round of powdery mildew. On the other hand, this could be the year you get a perfect balance of rainy and sunny days and everything flourishes.

Two, gardening is a skill and skills can be learned. You don't have to be born with a green thumb; you can learn to have one.

Some people acquire these skills so gradually that they seem more like instincts. Perhaps, unlike me, they did help Grandma in the garden while they were growing up, or they have simply learned by trial and error over many years of gardening. Although they may not be conscious of it, their gardening skills, such as knowing when to water and how much, come from careful observation and accumulated knowledge. That's good news, because even if you didn't grow up following Grandma around the garden, you can observe and you can learn what you need to know in order to become "good with plants."

You will need to pay attention to your plants and spend time on them, and the learning never stops. I don't think that's bad news at all. Paying attention, spending cocreative time in nature, and lifelong learning are just as important as putting food on the table.

Edible Sunlight

In these times, I find it comforting to remember that my true security—my survival and, more than that, my ability to thrive—doesn't come from my jobs, my paychecks, or my government but from the ground I am standing on. The plants that grow in that earth under my feet—and I try to remember that even under the pavement there is fertile soil—those plants harness sunlight through photosynthesis so that I can enjoy it as dinner. Because that is what food is: edible sunlight, given to us by plants. Even the animals we consume, if we eat meat, ate plants. This is really what keeps all of us alive. Without fertile soil, sunlight, water, and the plants that make use of those forces, there is no life as we know it on this planet.

During World War II, after more than a decade of economic depression, the government urged American citizens to plant victory gardens. The idea was that if people could produce more of their own food, it would free up resources that needed to be sent to the troops overseas and thus reduce the strain on the national budget. By 1943, twenty million Americans had planted victory gardens, many tearing up their front lawns in order to do so—including, by the way, Eleanor Roosevelt, who tore up part of the White House lawn to plant a victory garden, paving the way for Michelle Obama to break ground for an organic vegetable garden on the White House south lawn in 2009. During the World War II victory garden era, 40 percent of the produce in this country was grown in home gardens. Forty percent—just in case you were wondering how much of a difference individual gardens could make.

Although there are parallels, these are different times. But once again gardens are a path toward self-sufficiency. According to an article in the *San Francisco Chronicle,* nationwide "industry surveys show double-digit growth in the number of home gardeners this year [2009], and mail-order companies report such a tremendous demand that some have run out of seeds for basic vegetables such as onions, tomatoes, and peppers."

The home garden, and especially urban gardening, is an essential piece of the jigsaw puzzle of creating a sustainable food system. Our food has to grow somewhere, and with more than half the world's population living in urban environments and farmlands disappearing at an alarming rate, the question of where it is going to grow is a real concern.

If everyone in the New York City area chose to eat a mostly local foods diet, could the farms surrounding the city actually supply that much food? Probably not. But if we start to make use of backyards, community gardens, rooftops, windowsills, and terraces, they add up to a tremendous amount of space. Pair that with the surrounding rural areas, and then, yes, the NYC area could provide for itself, as could other metropolitan areas.

Jacquie Berger, the executive director of Just Food, says, "If city folks are going to grow their own food, they need to have a place to do it and they need to know how." Let's get to it.

Finding Your Garden

If your home is blessed with a garden, you can skip to the next section. If not, let's find you one.

Community Gardens

As of 2009, there are more than one million community gardens in the United States. Each community garden sets up its own rules for membership. In the two I've belonged to, each of us had our own plot in which we could grow whatever we wanted. We were also responsible for helping to maintain the communal areas, including helping out with the watering and weeding. To water, we had a permit that allowed us to hook up a hose to the fire hydrant on the street outside the garden.

Community gardens are more than a way for people whose homes are garden-less to get access to some growing space. They are also green spaces enjoyed by nonmembers. All Green Thumb gardens and most other community gardens are open to the public a minimum of ten hours per week. Many community gardens offer public programs, including cooking and gardening demonstrations, children's events, and food pantries. At the Garden of Union in Park Slope, people from the neighborhood can drop off their kitchen scraps in the composting bins at the back of the garden, and the local CSA gets to use the garden as its weekly distribution site. Some community gardens are big enough to function almost as small farms, and several have their own farmers' markets (see chapter 3, Sourcing Local Food).

Most community gardens have a sign posted on the front gate that gives you the contact info for how to get involved with the garden or get on their waiting list for a plot. You can also locate a community garden near you by visiting one of the Web sites listed in Useful Resources at the back of this book. If there isn't a community garden in your neighborhood, contact your community board. Often they have land set aside to meet public space requirements and will be delighted by the idea of using that space for a garden.

There are some terrific organizations that can help you start a community garden and provide support once it is established, including Green Thumb, Green Guerillas, Bronx Green-Up, Council of the Environment of New York City, and Brooklyn Greenbridge. Another great resource is Just Food's City Farms Toolkit. It is filled with tip sheets and other resources for garden organizing and planning.

Roof Gardens

As I mentioned, my first garden was a rooftop on West 26th Street in Manhattan. If you have access to your building's roof, you've got a terrific full-sun spot to grow food.

There are a few things to consider before you start hauling containers up there. First, check with your building's landlord or co-op board to get permission to install a roof garden. Find out if there are weight restrictions that you need to take into consideration. To help keep your garden within any weight restrictions, choose large but lightweight containers. The weight of the soil will keep large containers from blowing over, but small containers aren't stable on a windy roof. Don't plant anything near the street edge where containers could fall and hit someone on the head—yikes!

Next, consider how you will irrigate your garden. Container plantings dry out quickly on a hot, windy roof; and during summer's hottest weeks, you can count on needing to water every day.

I recommend choosing dwarf, compact, or bush varieties for your edible roof garden. They tend to be more suitable for container growing and don't need the tall trellises that can easily get blown over on a rooftop. The diminutive-sounding labels refer to the growth habit of the plants, not the size of the crop. You will still get a full-size cucumber from a bush variety just as you would from a standard long-vined plant, but it won't need trellising.

Front Yard Gardens

If your building has a front yard, or even space for a couple of large containers on either side of the door, consider planting something that is edible as well as ornamental. Juneberries *(Amelanchier sp.)*, for example, are covered with gorgeous white flowers in spring and bright-gold leaves in the fall. In June you can collect the blueberry-like fruit. If you are not the person responsible for your building's front plantings, find out who is and ask if they would be willing to plant something that is food for the mouth as well as for the eyes.

Front Steps

People often line the steps of the brownstone homes in my neighborhood with seasonal plantings of lovely flowers. They could be growing food there, and many food plants are also attractive. If you must have flowers, there are lots of edible ones that can enliven your salads as well as dress up your front steps.

In some neighborhoods where I've lived, if I put out a container planting in front of my building without securing it, it would be gone by the time I got home from work. If you live in such a neighborhood, you can chain your plant pots or boxes to the stair railings or fences.

Fire Escapes

It's technically illegal to put containers or anything else on a fire escape, but the reality is that many city folks do grow food there. If you choose to do so, remember that the fire escape's primary purpose is to provide residents (including you) with a means of getting out of the building in an emergency if the regular exits become

---

THRIFTY GARDENER'S TIP

If you take nutritional supplements, save those packets of polymer crystals that come in the jars to keep the capsules dry. They are the same substance sold as water crystals in garden centers for a steep price. If you mix them into the potting mix of your container plants they will soak up water and slowly release it over several days, greatly reducing how often you need to water.

---

inaccessible. Arrange your containers so they do not block access. It is also a good idea to secure them with wire to the railings so that there is no chance of them getting blown off the fire escape.

Windowsill Gardens

A lot of food can be grown in a window box, especially if is facing south or east and thus getting several hours of sunshine. If the box needs to provide eye candy as well as food, choose crops with interesting colors and leaf forms. For example, one window box planting I did included Peruvian purple peppers (the whole plant is purple, even the leaves), silvery sage, and red oak leaf lettuce. It looked gorgeous, and it produced food and seasonings for me from May till October.

Indoor Gardens

If you have a window that gets eight or more hours of sunlight a day, you can grow most vegetables indoors. Don't have a sunny window? Plant lights work well for many vegetables and herbs (more about that below).

Schoolyard Gardens

Alice Waters started the schoolyard garden programs on the West Coast; happily they are taking off here in the East as well. These gardens not only provide the kids with healthier food while reducing the carbon footprint of their school lunches but also teach the kids gardening skills that will be with them for a lifetime. They also learn where their food comes from. Nowadays many children (and more than a few adults) seem to think that the supermarket shelf is the source of their sustenance.

A few years ago I led a community garden event for junior high school students from the Bronx. Knowing that they were coming, I deliberately didn't harvest any tomatoes from my plot the week before. I wanted there to be plenty for my visitors. The tomatoes were hanging red and ready on the vines, but the students shuffled and stared at the ground and looked confused. Finally one of them asked me, how do we know which ones are ripe? I showed them how to tell, and then made a salad of the tomatoes they'd harvested plus some basil and other treats from the garden. At the first bite, the adult chaperones let me know that they were in taste bud heaven. But one of the boys said, "There's something wrong with these tomatoes; they're all juicy and wet." He'd never had anything but a mealy supermarket tomato.

Let's reintroduce our children to what a tomato can be. As Michelle Obama said in a recent *New York Times* piece, "When you grow something yourself and it's close and it's local, oftentimes it tastes really good. And when you're dealing with kids, for example, you want to get them to try that carrot. Well, if it tastes like a real carrot and it's really sweet, they're going to think that it's a piece of candy. So my kids are more inclined to try different vegetables if they're fresh and local and delicious."

Tree Pits

Don't do it. Although in recent years I have seen an increasing number of tomato plants and other vegetables planted under street trees, I don't recommend it. Plants can absorb heavy metals and other toxins, so harvesting from beside a heavily trafficked road is a bad idea. And then there are all those dogs whose owners regularly ignore the signs that say to please keep your animal away. There's a reason the plants on the sidewalk-side corners of tree pits usually look ragged compared with those on the street side. Need I say more?

However, as a forager and gleaner, I do harvest fruit and edible flowers from street trees. Fruits and flowers don't accumulate soil pollution as much as other plant parts. A man who lives a couple of blocks from me has a mulberry tree. He hates that the fruit stains the pavement every June and is more than happy to let me pick some of the berries. I spread a tarp under the tree and give the branches a good shake. Pints of delicious mulberries fall onto the tarp rather than onto the sidewalk. I'm happy and the neighbor is happy; it's a win-win situation.

If you have any say in what kinds of trees get planted on your block, why not choose ones that will feed you as well as shade you? (Read more about this in chapter 10, Feasting for Free.)

What Can You Grow?

Okay, now you've got a place to grow food. Before you make a list of what you'd like to grow, you need to consider what you *can* grow at your garden site. The main consideration is the amount of light your plants will get, since meeting the light requirements of your plants is the most important thing you can do when planning your garden.

Light

In order to produce edible sunlight, plants need—well, sunlight. Knowing the light requirements of what you want to grow is the key to getting a bountiful harvest. If you try to grow a full-sun plant such as eggplant in the shade, I guarantee you will be disappointed. Other plants are such efficient photosynthesizers that they can make do with less light, which is good news for shade-challenged gardens. Here are the light requirements for some of the foods a locavore can grow in our region. Let's start with the sun lovers, which include many of our most common vegetables:

Full-Sun Edible Plants
(Six to eight hours of direct sunlight a day)

- Broccoli
- Cauliflower
- Cucumbers
- Eggplant
- Mediterranean herbs. These include oregano, rosemary, thyme, basil, and marjoram. They are genetically adapted to growing on a rocky hillside with the Mediterranean sun blazing down on them, and those are the conditions you are trying to re-create when you grow them.
- Okra
- Peppers
- Potatoes
- Snap beans
- Sunchokes (aka Jerusalem artichokes)
- Tomatoes

Part-Sun Edible Plants
(Two to five hours of direct sunlight a day, but many of these plants also do well in full sun.)

Most of these are plants that would prefer full sun and are often labeled as full-sun plants. I've put them in the part-sun category in case you want to know what you can grow if you only have a few hours of direct sunlight each day. You can't grow these in full shade, but they will still produce a good crop on just a few hours of sunlight a day. How to know whether the plant you want to grow is in this semi–shade tolerant category? Just remember that with a couple of exceptions, leaf and root vegetables tend to be more tolerant of low-light situations than most fruits. In this case, I'm using the scientific definition of "fruit," which is "the seed-bearing part of the plant." If the part of the plant you intend to eat has seeds, it may need a lot of sunlight. What this means is that you can grow lettuce and beets on just a few hours of light a day, but don't try that eggplant.

- Asparagus*
- Blackberries and raspberries*
- Chervil
- Chives*
- Currants
- Horseradish*
- Hot peppers*
- Lemon balm (*Melissa officinalis*)
- Leafy greens (kale, bok choy, chard, spinach, lettuce, etc.)*
- Mint
- Parsley*
- Rhubarb*
- Root vegetables*
- Shiso (Perilla)
- Strawberries*

*You'll get a bigger crop if you grow these in full sun, but they will still produce well with just a few hours of light.

Now we get to the true low-light plants. Many urban gardens are shade challenged. Garden apartment backyards often feature a large tree, or there may be a tall building next door that blocks the light. In my own garden I have two elderberry shrubs that are over 10 feet tall. They are getting full sun and produce lots of fruit in August, but they shade out the raised bed in between them. Rather than giving up my elderberries, I've underplanted them with shade-loving edibles.

If your gardening space doesn't get direct sunlight, the good news is that there is still plenty of food you can grow. Woodland plants evolved under the shade of trees and actually prefer less light. These include some wonderful foods native to our northeastern forests, including the wild ginger and ostrich fern I mentioned earlier. (Ostrich ferns send up the "fiddleheads" that are a gourmet treat you sometimes see at the markets in early spring. They are usually expensive, so why not grow your own?) These woodland plants may be less familiar than supermarket ingredients, but consider that your invitation to get creative in the kitchen. (See chapter 8, Making Friends with your Kitchen.)

Part-Shade Edible Plants
(Dappled or bright indirect light)

- Elderberry (another shrub that will give you a bigger crop with more light, but fruits quite well with less)
- Mayapple (*Podophyllum peltatum,* a native plant with delicious fruit)
- Ostrich/Fiddlehead ferns (*Matteuccia struthiopteris*) There are a few other ferns including cinnamon fern (*Osmunda cinnamomea)* that produce fiddleheads, but in my experience ostrich fern is one of the tastiest and easiest to grow.
- Ramps/Wild leeks (*Allium tricoccum*)
- Spicebush/Appalachian allspice (*Lindera benzoin*)
- Sweet woodruff (*Galium odoratum,* the flavoring for German May wine. You can make your own by soaking the flowers and leaves of sweet woodruff in Riesling wine overnight.)
- Wild ginger (*Asarum canadense,* a native Northeast woodland groundcover; no relation to the commercially used Indonesian ginger, it has a similar—I think better—flavor.)
- Wintergreen (*Gaultheria procumbens*)

Full-Shade Edible Plants
There are not too many edibles that will grow in deep shade, but there are a few. Some are plants that most people grow as ornamentals without realizing that they are also edible, such as hostas. (Yes, you can eat them. When they first come up in spring and the leaves are still curled into a funnel shape, the Japanese use them as a

green vegetable in soups. They are quite good and will recover from your harvesting with impunity.) Plants for a Future (www.pfaf.org) has an extensive database that includes edibles for every light condition, including full shade. If you use their site to research possible plants for your edible landscape, keep in mind that they are in the U.K., which has a milder climate than ours. Look for plants that are hardy in Zone 6 or lower. (The zoning system is something gardeners use to categorize how cold- or heat-hardy a plant is. In North America, Zone 10 is Florida, the hottest; Zone 1 is way up in the Northwest Territories of Canada, the coldest).

There is one food that does show up in every cookbook and that will grow in the dark: mushrooms. These are fungi, not plants, but I include them here because you can grow them in full shade; some types will even grow in a closet. Mushroom-growing kits are available online from Fungi Perfecti and other companies (see Useful Resources).

Indoor Light Requirements

When plants are grown indoors, their light requirements change. Even a plant in the sunniest south-facing window will need more hours of light than it would outdoors because window glass filters the light spectrum. Raise the number of hours of light your plants need by two hours for each category of light requirement—e.g., eight to ten hours for full sun rather than six to eight. If using plant lights, plan on twelve to fourteen hours per day for full- and part-sun plants, and make sure that the plant light is no more than 2 feet above the tops of the plants.

Water

The next thing after light that you need to provide for your garden is irrigation. How will you water your edible garden? If it's just a window box, hand watering may do. But if your plants are in the ground, or if you have a large number of containers, watering is a major concern and can become a major chore. During the hottest weeks of summer, container plants may need to be watered every day.

The best solution, if you can afford it and if it is an option, is to install a drip irrigation system on a timer.

In my Brooklyn garden, that's not an option. The water faucet for the garden is located inside the neighbor's downstairs hallway, which makes hooking up a hose inconvenient and setting up an irrigation system nearly impossible. My solution has been to install a large rain barrel connected to the drainpipe that comes off the roof. One good rain is enough to fill the barrel, and it is easy dip my watering cans in as needed. I water a few areas of the garden each day, spreading out the workload over the week.

When hand watering, you want to make sure to water deeply so that the moisture reaches the entire root ball of the plant, not just the surface. Frequent shallow waterings are not as effective as occasional deep ones. With container plants, the time to water is when the top inch of soil is dry. The easiest way to check this is to stick a forefinger into the soil up to the first knuckle. Water your containers until water starts flowing through the drainage holes; this is the only way to ensure you've watered the whole root ball. Don't water again until that top inch of soil (but no more) is dry. This is true for your houseplants as well.

Earth boxes and "self-watering" containers work by having a reservoir of water that is gradually wicked into the soil. They can considerably reduce the time you spend watering. Commercially sold earth boxes tend to be expensive, but you can make your own quite cheaply. (See Useful Resources for Web sites that provide directions for making self-watering containers.)

Soil

Fertile soil is as important as sunlight and water to growing food successfully—just ask any farmer. It is also something that is disappearing fast. The soil on conventional farms is often depleted of nutrients and utterly dependent on chemical fertilizers. In addition, many of our richest soils have been paved over for highways, strip malls, and urban sprawl. Fortunately the earth is capable of regenerating to a remarkable degree.

The soil in those two community gardens where I had plots started out as rubble-filled dirt that couldn't grow much besides hardy weeds. But after a couple of years of intensive composting and organic gardening, the soil bounced back and became lovely rich loam in which almost anything I wanted to grow would flourish.

If you are starting a new garden from scratch in the city, my first recommendation is to have your soil tested for heavy metals and other contaminants. The County Extension office will tell you how and where to send a soil sample for testing. If you do get that bad news that your soil is too contaminated to grow food in, take heart. You can still garden on the site by building raised beds or planting in containers using purchased (and uncontaminated) soil mixes.

If your soil tests clean, the first thing you should do is set up a composting system somewhere. Mine is a black bin in the back corner of the garden. The New York City Compost Project sells these and other styles of compost bins at discounted prices. (See the composting section below for more about this important soil amendment.)

Your soil test may reveal that your soil is too acidic (most city soils are). You can remedy that by sprinkling garden lime and working it into the soil (not construction lime or the citrus fruit, but the gardening lime you can get at garden centers or online).

For containers, most commercial potting mixes are fine. I happen to like the Fafard brand. Be sure that you get a potting mix and not topsoil. Often they are right next to each other at the garden center. The topsoil is tempting because it is much cheaper, but it is meant to enrich in-the-ground gardens, not to go into containers. It is much denser with fewer ingredients added to improve drainage than potting mixes. If you use topsoil as potting soil, what you get when you water is mud that cakes, cracks, and pulls away from the side of the container as it dries. Stick to potting mixes.

Light, Water, Soil . . . and What Else?

Although light, water, and soil are the three most crucial elements of gardening success, there are other simple things you can do to help your plants thrive.

Container Plants and Drainage

If you are growing your food in containers, the containers must have drainage holes—no exceptions—because a sure way to kill your plants is to drown the roots by having them sit in soggy soil. If esthetics are less important than cost, you can make a plant container out of pretty much anything. I've seen old plastic buckets doing a dandy job of housing tomato plants. Just be sure to punch out or drill a few drainage holes. Wet potting soil can clog drainage holes so that they don't function. To prevent this, cover the holes with landscaping cloth, broken bits of terra-cotta pots, stones, or Styrofoam packing peanuts. This covering layer does not replace the drainage holes but protects them.

Composting

Compost is nature's own method of restoring nutrients to the soil. In the wild, leaves and other organic debris land on the ground and decompose with the help of time, worms, fungi, and beneficial insects. In the garden you can re-create and speed up this natural process by setting up a compost pile.

What can go into your compost? All vegetable and fruit scraps, eggshells, used coffee filters, coffee grounds, tea leaves, shredded newspaper, sawdust (if it's chemical free), and shredded egg cartons. What shouldn't go in? I don't compost meat or bone scraps because they lure the stray cats in my neighborhood to break into my compost bin (yes, they've figured out how to take the lid off). I also don't compost fruit pits or anything greasy because of how long it takes them to break down (years, maybe even decades).

Compost requires a mix of nitrogen-rich green ingredients and carbon-rich brown ones. The "green" ones include not only literally green materials, such as plants you've weeded out of your garden, but also coffee grounds and kitchen vegetable and fruit scraps that may or may not be literally green. The brown component can come from the fallen leaves of deciduous trees, chemical-free sawdust, newspaper, and egg cartons, among other things.

The minimum size for a successful compost pile is 3 square feet. At this size (or larger), the compost heats up in the center, speeding decomposition and killing weed seeds. The compost should get some air (commercial bins usually have some holes in their sides). Turning the compost, or transferring it occasionally from one pile or bin to another, speeds up the composting process.

If you have space for a compost bin in your garden, get or make one. If you don't, try an indoor earthworm composting system. I did one of these in a closet for a few years until I had an outdoor garden. It is not nearly as messy as it sounds. Basically you set up a covered plastic bin and put in damp, shredded newspaper and red worms. Then you feed the worms your kitchen scraps, including vegetable peelings, eggshells, apple cores, etc. In just a few weeks you can scoop out some fertile, dark compost to add to your potting mixes or use in your community garden plot. And no, it doesn't smell. In a worm composting system, the damp newspaper provides the "brown" and your kitchen scraps the "green." The New York City Compost Project sells worm-composting kits (including the worms) and also offers worm composting "worm shops."

Feeding, aka Fertilizing

Compost provides most of the nutrients in-the-ground plants need, but many vegetables are heavy feeders (meaning they need lots of nitrogen and other elements present in the soil) and will require some supplemental feeding. This is especially true if they are grown in containers. With container plants, you are all the nature that plant is ever going to know. You decide what kind of earth it will grow in (potting mix), how much sunlight it will get to photosynthesize (depending on where you put the container), and how much "rain" (your watering system) it will drink. The nutrients in the potting mix in your containers are used up by your plants and not replenished by natural systems, such as slowly decomposing leaves.

The soils industrial crops are grown in are so depleted at this point that they literally can't supply the needs of the plants. In addition, the (often genetically modified) varieties grown are bred to need a heavy dose of chemical fertilizers. If you planted corn seeds in such soil and didn't dump chemicals on it, chances are you wouldn't get much of a crop.

Organic gardening actually helps rebuild the fertility of the soil its crops are grown in. It doesn't make sense to buy organic at the store and then douse your bell peppers with Miracle-Gro. One of my favorite organic fertilizers and soil conditioners is seaweed. This is available in both liquid and granular forms. Fish emulsion is also excellent, but I can't use it in my neighborhood because it drives the stray cats crazy. My favorite source for organic fertilizers (besides my own compost bin) is listed in the Useful Resources section.

Horse manure is a classic farm fertilizer and is available for free from many of the city's stables and the mounted police. You just have to contact them in advance and show up with something to carry the manure away in. (Plastic tubs with tight fitting lids work well. A less-well-sealed container is a guaranteed way to get extra seats on the subway.) Never put unaged manure directly into the garden, as it can burn the plants. Mix it with your compost instead and let it age for several months before using it as excellent fertilizer.

Mulch

Spreading a layer of mulch around your plants helps keep down weeds. It also keeps the soil from losing moisture quickly during hot weather. In winter it protects the roots of your perennials (plants that will come back year after year) from the extremes of freezing and thawing soil. As mulch decomposes, it improves the soil, mimicking nature's own forest floor composting system.

Commercially available mulch is usually made from bark that is either shredded or in chips. These mulches do the job well and keep your garden looking tidy, but there are other options. Bark mulches needs to be spread at least 3 inches thick in order to be effective. A layer of mulch needs to be porous enough to let water through but thick enough that it will block weed growth.

Leave 2 to 4 inches of space between the mulch and the base of the plants you are mulching around.

Cardboard or newspaper can be laid on the ground around your plants and then weighted with rocks, compost, or grass clippings to keep them from blowing away. If using newspaper, put down a layer at least ten sheets thick to effectively control weeds.

The one downside to tree-based mulches (bark, newspaper) is that as they break down they can rob the soil of nitrogen. You can get some nitrogen back into your soil by adding an occasional top dressing of grass clippings (not always easy to find in the city, but if you have them . . .), compost, or kelp granules (available at garden centers and online; see the Useful Resources appendix).

There are also plastic mulches, but these are petroleum products, so I avoid them.

Pest and Disease Control

Harmful insects and diseases will show up on some of your plants at some point, no matter how attentive a gardener you are. Don't beat yourself up about it.

The first line of defense is buying clean plants. Inspect any plant you are bringing home to your garden very carefully, looking at the undersides as well as tops of the leaves. If the plant doesn't look robust, or if anything flies away when you shake it, don't bring it home.

Once the plants are in your garden or kitchen window box, pay attention. If you think a plant doesn't look as healthy as it did last week, if the leaves look mottled or are curling up oddly, don't assume it's just a question of whether the plant needs more water. Close examination of the plant will most likely reveal bugs or other problems, and the best solution is to catch the problem early on, while it is still manageable.

Caught early enough, simply removing a few infested or otherwise less than lovely looking leaves can be enough to curtail the problem. In my experience, fungal diseases such as powdery mildew and black spot don't really respond to any of the organic solutions available. The best action is to remove the affected leaves. Never compost plants that were diseased or infested.

Soap spray will take care of many insect problems. There are horticultural soap sprays available, but I find that a half teaspoon of Murphy's Oil Soap in a pint of water in a spray bottle works just as well. Be sure to spray the undersides of leaves and growth tips—that is where insects tend to congregate.

There are also beneficial insects you can introduce into your garden that will eat the bad guys. (In the garden, vegetarians are evil because they eat your plants; carnivorous bugs are good because they eat the bugs that eat your plants.) Ladybugs and lacewings are among the beneficial insects you can buy and set free in your garden.

Not sure which bug is eating your plant or what that brown spot on your plant's leaves means? An excellent diagnostic catalog of photographs is available online at www.gardensalive.com. (They will of course try to sell you their products, which are excellent but optional.)

Slugs and snails are likely to find their way even up to rooftop gardens. The simplest and most effective solution is beer. Slugs and snails can't resist beer. Set out shallow bowls of it wherever you suspect slugs and snails (the leaves of your plants will have holes in them as if something has been munching on them, which it has). These critters come out at night. In the morning you will have a bowl of slug brew. Throw it into the compost.

Weeding

There is no such thing, scientifically speaking, as a weed. A weed is just a plant that you didn't introduce into your garden and don't want growing there. To weed or not to weed is a question of what is in fashion at the time. Dandelions, for example, were intentionally introduced to North America by Europeans as a source of food and medicine. Now they are considered the bane of lawn tenders nationwide.

No matter how potentially valuable they may be, those "volunteer" plants that pop up in your garden are competing for valuable resources with the plants you have chosen to grow. They are stealing water, sunlight, and nutrients from your crops. If you want your crops to thrive, the weeds have to go.

In my own garden I practice something I call selective weeding. Because so many of the "weeds" that appear in my garden are wonderful wild vegetables, I often let them grow until they are big enough to be worth harvesting. After that, though, I am merciless. Even the edible weeds must make way for my tomatoes, basil, asparagus, and raspberries. The nonedible weeds get yanked out as soon as I spot them. (To learn more about which weeds are safe to eat and which are not, see chapter 10, Feasting for Free.)

The easiest time to weed is the day after it rains. The plants pull out easily then. With small weeds that have just come up, you can simply scrape the surface of the soil to uproot them. In a closely planted small garden, my favorite hand tool for that job is called a CobraHead. It can get at the weeds in between my cultivated plants without accidentally uprooting the plants I want to keep.

What's Worth Growing

Now that you know what you can grow given the amount of light and other resources you can provide your plants, it's time to make some choices. If you have a big full-sun garden, you can probably grow most of your own vegetables. On the other hand, if space is limited you may want to grow just those that tend to be expensive at the farmers' markets or varieties that are hard to find.

What you want from your edible garden will partially depend on where else you get your food. For example, my CSA share provides me with an abundance of leafy greens, root vegetables, squash, and snap beans, so although I could grow these in my garden, I don't bother. What I do grow are certain heirloom tomatoes I love and don't get from my CSA—the ones that are always priced high at the farmers' markets, such as Black Krim, Brandywine, and Purple Cherokee. I also grow asparagus, red currants, elderberries, and rhubarb because we never get those in our CSA. I grow every culinary herb I can get my hands on because, although we do get fresh herbs as part of our CSA share, we never know which herbs we will get from week to week, and I like to have a full range of fresh herbs at my kitchen door.

So, if you have the room, light, and time to grow your own vegetables, skip the CSA share and go for it. If you get some of your food from a CSA or farmers' market, then grow the ingredients you love that would otherwise be costly or hard to find.

You'll find great sources for plants and seeds in the Useful Resources section.

What to Plant When

Although it is cheaper to plant seeds than to purchase plants for your garden, in the New York City area our growing season is too short to start some vegetables from seed outdoors. If you have enough space and direct sunlight (or plant lights) in your home, you can start plants indoors in March and transplant them outside in May when the weather is warm enough. Don't be fooled by balmy afternoons in April: What matters to the plants is not daytime but nighttime temperatures. Until nights are reliably above 50 degrees, it is still too early for tomatoes, peppers, eggplants, and cucumbers to go outside. But other plants are more cold tolerant. Here is a brief guide to when you can plant some common vegetables and herbs in our area:

Indoors from Seed in Late March (Transplant outdoors in May)

- Chili and sweet peppers
- Eggplant
- Tomatoes
- Basil (*Note:* Aside from basil, cilantro, shiso, and chervil, I recommend purchasing most herbs as plants rather than starting from seed. Perennial herbs such as sage and rosemary germinate and grow slowly until they are well established.)

Outdoors from Seed in Mid- to Late April

- Beets
- Bok choy
- Carrots
- Chard
- Kale
- Lettuce
- Mustard greens
- Radishes
- Spinach
- Turnips
- Cold-tolerant herbs including calendula, chervil, parsley, and shiso

Outdoors from Seed in May

- Beans
- Corn
- Cucumbers
- Summer and winter squash

The Neighbors: Joys and Oddities of Urban Gardening

Urban gardening can produce some interesting neighbors. Because city buildings are so close together and open lots such as community gardens are right on the street, privacy is not necessarily a given in the urban garden. I share my backyard with the folks in the apartment next to mine, but my garden is also in plain view of the windows of all the buildings surrounding it and the gardeners on the other side of the fence.

Machete Woman became my friend when she realized that I was growing food.

Her subsistence plot is right next door to my garden. I call it a subsistence plot because there is no eye candy in it, nothing grown purely for beauty. There is a fig tree. There are old mops sticking out of the ground, mop end up. She uses them for tomato stakes. The tomato plants are tied to the mop handles with strips of nylon stockings. Beans, peppers, collards, and cucumbers make up the rest of the garden design. In winter she wraps old blankets and bathmats around the base of the fig tree to protect it.

When I first moved to this apartment, I took note of my neighbor's rudimentary but effective gardening style with interest, but I never saw anyone out there gardening. Meanwhile I was busy getting my own garden started. I thought that the fence between our two gardens was calling out for vines: morning glories for something pretty maybe, or something more useful such as grapes or hops? I opted for hops, knowing that they grow and climb voraciously and would be quite capable of covering the fence in a single season. They also produce delicious, asparagus-like shoots in early spring, and in early fall the strobiles (the technical term for hops "flowers") can be used to flavor beer and make an herbal sedative. Edible, medicinal, and attractive, I thought they'd be perfect for the fence.

I was baffled when weeks passed and the hops vines made no headway. They got longer but every morning I'd come out and they'd be in a sorry little pile on the ground, not climbing and twining upwards the way I expected them to. The mystery was solved one morning when I came outside and saw a middle-aged woman carefully reaching through the spaces in the fence to untwine my hops vines. She didn't notice me. "Don't like nothing on the fence," she muttered, "nothing on the fence, nothing."

I called "hello," but she didn't respond. Her head was wrapped in a bandana and she was wearing what my grandmother would have called "a kitchen dress," one of those nondescript floral print garments that could as easily have been a nightgown. Beside her feet was a shiny, obviously well cared for machete. Machete trumps garden design plans. I moved the hops vines.

That machete seemed to be her sole gardening tool. I watched her take a wide-legged stance with the machete raised over her head. Thwack! The machete dove into the ground. Then she jimmied it back and forth; when the hole thus created was wide and deep enough, she planted a bell pepper seedling in it.

She didn't speak to me much for several more weeks. I grow a mix of edibles, medicinals, and ornamentals, and my food plants are tucked in among the others. I suspect she thought I was just another dabbler wasting precious garden space on stuff you can't eat. Finally one day she responded to my greeting. She gave me a disapproving look and then her eyes returned to the ground. She asked if I was going to grow any food. I replied that yes, I was growing food. I pointed out the vegetables and herbs I'd already put in, tucked among the other plants. Her face softened and she looked up and met my eyes with a hint of a smile. "I'm Cameron," she said.

We've become friends, and each fall she generously shares some of the figs from her fig tree. I give her some of my basil, mint, and other herbs, which she doesn't grow but loves in recipes. I guess she doesn't have to grow them since I grow them for her, just like I don't need to plant a fig tree.

Even in the garden, nothing brings people together like food.

IF YOU DO JUST ONE THING...

Grow some herbs indoors.

Growing your own herbs indoors lets you snip off just what you need for a recipe, rather than buying a whole bunch that is way more than you need. I like to keep some herbs going indoors even though I am lucky enough to have a garden. When it's pouring rain outside, do I really want to go out and pick some chives? No. But since there are some growing in my kitchen window, I don't have to.

To grow herbs indoors you need a window that gets at least six hours of sunlight (more is better), or a plant light. If you are using a plant light, make sure you position it no more than 2 feet above the tops of the plants and that you leave it on for twelve to fourteen hours each day. An easy way to do that is to put the light on a timer.

Plant your herbs in a container at least 6 inches deep. Make sure your container has drainage holes. Any commercial potting mix will do for most herbs, but be sure to use a potting mix and not topsoil. To further improve the drainage (important for herbs), you can add 25 percent perlite or vermiculite to the potting mix.

A few of the best herbs for growing indoors are chives, oregano, parsley, mint, rosemary, cilantro, and sage.

6
The Cost Factor

"But it's so expensive!"

This is many people's first objection to eating local and organic foods. Typical consumers suffer sticker shock when they shop for organic food and turn away discouraged from $4-a-pound tomatoes at the farmers' market. Maybe they are just used to the lower prices at the mega-mart, or maybe they're trying to come up with next month's rent and the 99-cent burgers at Mickey D's are looking like a great deal compared with the sustainably raised local beef.

The stereotype of someone who eats a local, organic diet is a foodie with a disposable income and time to spare, someone who can afford to think about things like how their food choices will affect the environment because the difference between farmers' market prices and Costco isn't going to make or break them. Yet unless a significant number of people at all income levels have access to locally grown, organic ingredients, sustainable agriculture isn't going to have enough of an impact on the environment or on our health to make a positive difference.

The good news is that there are ways to eat fresh, sustainably grown local foods no matter what kind of a budget you're on. For example, from June through November I get my locally grown fruits and vegetables for free.

The True Cost of Food

Why is the cost of local, organic foods often higher than average supermarket prices? Because it is based on the true cost of producing the food, unaided by government subsidies of commodity crops, "cheap" oil, and underpaid labor. We pay more for our food than we realize because at tax time we pay for those government subsidies. Small family farms rarely get subsidies, and they pay their workers a living wage. If the farm is organic, then as Farmer Ted (my CSA farmer) says with a sigh, the solutions to pests, diseases, and weeds are more costly than the pesticides used in conventional farming. Industrial agriculture has not only had a destructive effect on the environment, it has skewed our perception of what food should cost.

There is also a shortage of fertile, affordable land. Ideally an organic farmer has enough land to plant half of it with that year's crops and let the other half lie fallow under a cover crop. The cover crop eventually gets plowed into the soil, adding nitrogen and other nutrients, and those freshly revitalized fields are planted with vegetable crops the following year. This alternation enriches the soil better than merely composting a piece of land that gets intensively planted with crops year after year. It also cuts down on weeds, which are a big source of expense and labor for the farmer.

"That's how organic farmers really can take good care of their soil," says Farmer Ted, "and when you grow cover crops the weed pressure is greatly reduced." But even though he would love to farm that way, he isn't able to yet because he simply doesn't have enough land. He needs to plant every inch of what he's got (fifty acres total, some of which are rented) with food crops in order to make ends meet.

The land issue is a big one for animal farmers too. In order for animals to be raised humanely, fed by self-renewing pastures, and pastured in a way that the waste created by the animals can be reabsorbed into the land as fertilizer rather than toxic overload, the animals need space. But the farmer needs to raise enough animals to meet demand and pay the mortgage. And good, affordable land is hard to find.

Those are some of the reasons the price tags on local and organic food are often higher than those produced by conventional farming. But the tide is turning. There are an increasing number of programs working to make locally raised food affordable to low-income neighborhoods while still paying the farmer a fair price. (See Useful Resources for a list of these organizations.)

The rest of this chapter focuses on ways to make eating a local foods diet affordable, regardless of your income level. We'll start with free and progress to cheap. With that said, if you can afford it, I urge you to support the local farmers whose specialty products may cost a bit more. In addition to getting delicious, healthy food, you'll be voting with your dollars for a clean environment, a thriving local economy, and fair wages. You'll also be helping to keep those farmers on their land.

Following are the best ways I've discovered to eat local on a budget.

Free Local Food

Become a CSA Core Member

If you can volunteer a few hours a week for a few months each year, you may be able to get a community supported agriculture (CSA) share totally free. CSAs depend on a core group of volunteers who are responsible for tasks that range from bookkeeping to Web site maintenance to communicating with the farmer. In exchange, most CSAs offer core members discounted or free vegetable shares, depending on how much time they put in.

My own core member assignment is being the site coordinator for the Park Slope CSA. Most Tuesday afternoons from June through November, I am at a community garden in Brooklyn setting up tents and hauling bins of food. The garden is our distribution site. My responsibilities include organizing each week's teams of volunteers, getting the fruits and vegetables off the farmer's truck, and setting up the site so that everything is ready when the members come to pick up their weekly shares. Our CSA asks only that someone volunteer eight weeks as site coordinator in exchange for a free vegetable share. Because I put in at least double that, I also get a free fruit share. And because I do a lot of food preserving, I am able to make this free bounty extend long after the weekly deliveries stop for the winter (see chapter 9, Simple Food Preservation).

Along with reaping the benefits of free food, I enjoy being site coordinator. Yes, it's a lot of work and a lot of time out of my hectic life, but it's also fun. I get to chat with the farmer and to meet all the other CSA members. We exchange recipes, get excited when the first tomatoes appear, and talk about what we'd like our farmer to grow more of next year.

Sign Up for the Late Shift

There's another way to get free food from your CSA (and I hope I don't get in trouble for sharing this tip). In addition to all the hours core members put in, most CSAs require lesser time commitments from non-core members. In our CSA, for example, everybody has to do two shifts of a couple of hours each at some point during the agricultural season. Each member gets to choose when he or she want to work those shifts.

Imagine it's the end of a weekly distribution. It's autumn, so this week's share included beets, lettuce, carrots, garlic, kale, celeriac, apples, butternut squash, potatoes, and leeks. Seven members didn't pick up because they were on vacation, and several people didn't take their beets because they don't like them. So there are extras of everything, especially beets. In a few minutes the driver from City Harvest will arrive and bag up what's left for a local soup kitchen. But first, those of us who are volunteering for the last hours of the weekly distribution get to help ourselves to whatever extras we think we will use. (I'm looking forward to canning some beets!)

So when you're signing up for your CSA work assignments consider choosing the late shift. There's no guarantee, but you may luck into some extra food for free. Please only do this if you can actually use the extra food. It would be a shame to throw food out when it could have gone to food pantries and soup kitchens via City Harvest.

Food Pantries and Soup Kitchens

What happens to those extras at the end of a CSA distribution if the members don't take them? They go to food pantries and soup kitchens, as do many of the leftovers at the end of the day at the farmers' markets. United Way works with Just Food, City Harvest, and other organizations to arrange for the food to be picked up and taken to approximately 370 emergency relief organizations that provide food to low-income and homeless families in New York City. They offer grants that make it affordable for CSA farmers to give this food to the relief organizations; they also share information and technical assistance to help those organizations operate. Persons eating food pantry and soup kitchen meals may be eating local without realizing it.

Wild Edible Plants

There are delicious wild fruits and vegetables as well as gourmet mushrooms growing all over the city, not just in the parks but likely on your doorstep. Usually overlooked as "weeds," these free foods are yours once you learn some simple but essential identification skills. Numerous tours and classes on wild edible plants are available in the city. (See chapter 10, Feasting for Free, to learn more about safely identifying and harvesting the wild edible plants that grow near you.)

Barter

On a recent working trip to Switzerland (not remotely local, I know), I stayed in an apartment in a neighborhood filled with gardens. My landlady, Frau Arnold, lived upstairs. She had several different jobs, one of which was creating lush professional flower arrangements. One afternoon we arrived back at the house at the same time. She had a big bag of produce with thick, red rhubarb stalks sticking out of the top. I complimented her on how good they looked, and she promptly took two of the rhubarb stalks out of the bag and gave them to me. Then she ran upstairs and came back with a package of sugar "so you can make compote." (I would have preferred local honey, but I didn't want to be an ungrateful locavore, so I accepted the sugar.) Then she explained to me that the rhubarb and other produce in the bag all came from a neighbor's garden. For years she's had a barter arrangement with the neighbor. She gets fruits and vegetables from the garden, and the neighbor gets a weekly flower arrangement from Frau Arnold.

Barter is alive and well in the twenty-first century, and if you know someone with a garden, you may be able to trade your own service or product for their surplus. Maybe those friends who spend weekends in the country need you to feed their cat while they're away. Or maybe someone you know has a plot in a community garden with a zucchini plant that is overproducing, as zucchini are infamous for doing. Well, okay, they'd probably just give you the zucchini. But you get my point.

If you have a skill, or time to babysit, clean, run an errand, or pet-sit for someone who has a garden prolific enough to supply extras, you can get some of your food without laying any cash on the table.

The rhubarb compote was excellent, by the way.

Don't Throw That Away! Food Scraps and Parts You Might Not Be Using,
But Should

Instead of throwing apple cores and peels into the compost, I stockpile them in the freezer to make homemade apple vinegar and to use as pectin for jellies and jams. I also use my freezer to save vegetable trimmings and poultry, meat, and fish bones, which I turn into delicious stocks that later become soups and sauces. (See chapter 8, Making Friends with Your Kitchen, for recipes that turn what many people throw away into wonderful pantry ingredients.)

Blue Moon Fishery and other seafood mongers at the greenmarkets give away free bones and scraps for stock during winter. You just have to ask. This brings up the whole something-for-nothing concept of stock. Yeah, I know those Food Network chefs are telling you to go out and buy a fortune of ingredients to go into your stock. Ignore them. Start a bag or container in your freezer dedicated to stock. If you are a vegetarian, you'll need only one. If you eat fish, poultry, or meat, plan on a separate bag for each. For the animal products, when you have bones, throw them into the bag after the meal. Here are some other things you can save for stock:

- Parsley stems
- Heel ends of onions after you've chopped up the rest to cook with (Cut off the fibrous roots.)
- Carrot ends (well washed) and leaves
- Soaking water from rehydrated dried mushrooms or tomatoes
- Stem ends of tomatoes after you slice them off (There's usually still a good bit of tomato attached, which is the part you want for your stock.)
- Green parts of leeks (well washed)
- Celery stalk ends and leaves

Many root vegetables have edible, delicious leaves. Most of these are best cooked, but some can also be added to salads. Root vegetables with edible leaves include radishes, turnips, carrots (strong flavor—use these as a fennel-like seasoning), and beets. (Beets are the same species as chard, and the leaves can be used identically.)

Cheap/Discounted Local Food

Buy It When It's Cheap (and Good!)

When the first tomatoes appeared at our local farmers' market this June, they were $4 per pound. Ouch! Fortunately I knew that it wouldn't really be peak tomato season until late August, so I waited. Sure enough, prices dropped to $1.50 per pound as tomatoes came into peak season.

Eating seasonally is crucial on a local foods diet, and it is one of the simplest ways to keep costs down. Each crop has a season and a peak season. Peak season is when the produce is at it's best and also when the prices drop. (For more on what's in season when, see chapter 4, Eating with the Seasons.)

Don't Waste Food

Pinching pennies with groceries doesn't make sense when you throw out half a head of lettuce or most of a bunch of fresh basil because you couldn't use it before it spoiled. Yes, it could go into the compost, but that would be some pricey compost. Learning how to store produce so that it lasts, and what to do with a surplus, means not wasting food or money.

For example, yesterday I picked up my weekly CSA share. We got carrots again. I still haven't used up the carrots from last week. Fortunately this week's CSA newsletter included tips on the best ways to store the produce we're getting.

One of the best tips is simple: Use the most perishable stuff first. This includes leafy vegetables like lettuce, herbs, and soft fruits like berries. Hard vegetables will keep for weeks or even a month. These include root vegetables such as beets and turnips, as well as winter squashes and onions.

Any produce you store in the refrigerator will keep longer if you invest in some "green bags" (available at some supermarkets and food co-ops or online at www .greenbags.com). The way they work is that they release the natural gasses that fresh produce gives off. In regular storage, these gases cause fruits and vegetables to either ripen or rot. (This is the science behind putting a ripe apple in with under-ripe tomatoes or avocados to get them to ripen.) By harmlessly sequestering the scentless natural gasses away from the produce, your food stays fresh for days or weeks longer than it would otherwise.

If you buy your fresh herbs rather than growing them yourself, usually you bring home a big beautiful bunch that is way more than you can use before the leaves start to wilt. Some herbs, especially sage, oregano, and thyme, dry well (see chapter 9, Simple Food Preservation), but others like cilantro and parsley don't. For herbs that don't dry well, put 1 to 2 inches of water in a glass or jar and place your bouquet of fresh herbs in it. Cover with a plastic bag and store in the refrigerator. Even delicate herbs such as cilantro can last as long as two weeks stored this way. *Tip:* Wash the herbs before you put them in the glass of water so that you can just break off what you need without the fiddly task of washing a few individual sprigs. Dry the leaves with a kitchen towel before covering them with the plastic bag.

Put Some By for the Cold Time

When I first started my 250-mile diet, I was seriously concerned that my food budget was going to start eating into my rent money. I am not someone with disposable income to spare. Because I was also worried about what I would eat in winter, I wanted to put up as much food as possible. So in addition to what I needed for immediate nutrition, I was buying lots of extra fruits and vegetables to dry, freeze, and can. I did have the sense to stock up only on what was in peak season and therefore cheapest, but it was still costing me more than what I'd spent on food in the past. When winter came, I found I didn't need to buy much of anything at all because I had put up so much food in summer and fall. My winter food budget was drastically less than what I used to spend. I didn't keep a strict tally, but I didn't dip into the rent money. And I estimate that by the end of the year, I came out even with what I spent on food in previous years.

Even if you ate only what was at its in-season best and cheapest all summer, you could still get slammed when winter comes unless you practice a little food preservation. It's not just that out-of-season food is more expensive. If you live in a cold-winter climate, you may find that there is little or no fresh local food available in February. Or if there is, it will be hothouse-raised vegetables, usually very expensive and also not as environmentally friendly as other local foods because hothouses require fuel to heat them. The non-hothouse late-winter fare here in the Northeast is admittedly lacking in variety: mainly apples, potatoes, apples, cabbage—and did I mention apples?

But with a little food put up for the cold time, my locavore meals in winter are varied and delicious, and they balance the budget of what I spend during the warm months. The strawberries I froze when they were at their most luscious in June become breakfast smoothies in January; the ratatouille made with summer squash and eggplant becomes a quick pasta sauce long after squash and eggplant season is over.

It's well worth learning a few food preservation skills, not only to add interest and nutrition to your winter diet but also to keep costs down. Many people are daunted by the idea of preserving food, but it can be as simple as tossing a bag of berries in the freezer. To get started, see chapter 9, Simple Food Preservation.

Get to the Farmers' Market Late
The best time of the day to get to the farmers' market if you are trying to save money is just before they pack up for the day. True, the farmers may have run out of some

things and there may be less to choose from than there was first thing in the morning. But at the end of the day, the farmers often start marking down prices in hopes of moving more produce before making the trip back to the farm. Keep an eye out for bags of imperfect but still completely acceptable food. At the Union Square market in Manhattan, I've often picked up big bags of onions, carrots, or potatoes for a dollar; tomatoes for half what they were earlier in the day. The vegetables may have a blemish or two, but they'll be fine for sauces and salsas.

Join a Food Co-op

I belong to the Park Slope Food Co-op. Not only does it have a good selection of local produce during the growing season, it also carries other foods and household items that a regular grocery store would carry. Everything at the co-op is 20 to 40 percent cheaper than its counterpart at the supermarket. The reason is that the co-op is member owned and not seeking to make a profit beyond what is needed to pay full-time staff members. There are very few full-time employees, because every member is required to work a two- to three-hour shift every four weeks. That worker restocking the yogurt is an unpaid member doing her shift. Ditto the person mopping the stairs (that would be me on the Sunday 7:00 a.m. "A" squad shift), as well as the guy taking boxes of okra off the truck. Although I confess I don't love mopping floors at 7:00 a.m. once a month, I know that as a member I benefit from the hours I put in. Not having to pay workers is a big part of what keeps the co-op's prices low.

Other co-ops operate differently. Some do not require members to work hours but do require an annual fee in exchange for the discounted prices. The annual fee is low enough to more than pay for itself if you shop at the co-op regularly. Some require an initial investment fee but return it to you if you leave the co-op.

See Useful Resources for Web sites that will help you find a co-op in your area.

Ask about Discounted Low-Income CSA Shares

Many CSAs offer discounted shares to low-income families. The difference in price is often paid for by donations from other members who can afford to put in a little extra. Most CSAs also accept EBT payments at a discounted membership rate. Ask if discounted shares are available at your local CSA, and be prepared to show some proof of your income status to qualify.

Start a Garden . . . or at Least a Window Box

You may be wondering why I didn't put this tip in the free-food category. After all, no one is going to charge you for that basil you just picked out of your window box. But gardening does include some start-up costs: tools, soil amendments, seeds (unless you save your own), plants, etc. Gardening is not a totally free enterprise. But it is the ultimate local foods diet, matched only by foraging for wild edible plants. (See chapter 5, The Zero Miles Diet, to get started on growing your own food.)

Eat Fewer Animal Foods

If you're a vegan, you can skip this and the next two sections.

Even sustainably raised animals and animal products such as eggs require a heftier input of resources and labor than plants do. In industrial agriculture, this is escalated to horrible extremes of thousands of acres devoted just to raising plants for animal feed, plus extensive environmental damage, cruel conditions, rampant livestock disease, and antibiotics and growth hormones you really don't want to be eating. (If you want the details on how most commercial animals are raised, visit www.themeatrix.com.)

Small-scale, pastured animal farms also have additional costs and labor above those required for vegetables. A head of lettuce just needs to be cut from its roots and packed onto the farm truck. But meat requires butchering and packaging. Milk has to be pasteurized and bottled, cheeses made and aged before they reach you. Eggs have to be sorted and packed into cartons, and the chickens probably need winter feed grown or purchased specifically for them even if they were pastured during the summer. Hence the higher costs for animal products.

In many traditional cultures, meat is used more as a condiment than as the biggest thing on the dinner plate, with the exception of special occasions. It is one thing to enjoy a roast chicken with your family on Sunday, quite another to wolf down a bucket of chicken every few days, with time out for hamburgers and hot dogs in between.

By eating vegan meals several times a week, even if you enjoy your cheese or meat on the other days, you'll significantly reduce your food costs.

But If You *Are* Eating Meat . . .

One burger becomes two burgers. Any ground meat can be turned into meat loaf or meat loaf patties ("burgers") by adding finely chopped, sautéed vegetables, seasonings, a beaten egg and some bread crumbs.

Use the whole bird. Per pound, chicken is cheaper when you buy the whole bird and cut it into parts yourself. (The *Joy of Cooking* has excellent instructions on how to do this.) After you've separated the wings, breasts, legs, and thighs, what you'll have left is the back. This makes excellent stock.

Eat the offal. Before you turn up your nose at this idea, let me urge the omnivores who are reading this to learn a few delicious recipes for serving up "the nasty bits." If the mention of livers, kidneys, tongues, and hearts makes you cringe, then please tell me what part of killing an animal and then not using the whole animal do you think is a good idea? Organ meats are cheap, often less than half the cost of other meats, and loaded with more vitamins and minerals than their muscle-meat counterparts. They are also very tasty if prepared correctly.

At my CSA meat farmer's suggestion, I've made some excellent recipes from offal, including a homemade liverwurst that beat anything from the deli. A great book devoted to recipes that make use of "the nasty bits" is *The Whole Beast: Nose to Tail Eating* by Fergus Henderson. If you're not opposed to the idea but wouldn't know what to do with these cheap but healthy meats, there's an all-purpose recipe for them in chapter 8, Making Friends with Your Kitchen, that you and guests will gobble up (your call whether or not to tell them what it is).

Take advantage of free cooking fats. Okay, here's another one in the make-use-of-the-whole-animal category for omnivores: Save the rendered fat from bacon and poultry. This is easy to do. For bacon, once it's cooked and out of the pan, pour the drippings through a paper or cloth coffee filter into a glass jar. For poultry I trim off the fatty bits of skin before cooking and store them in the freezer until I have about a pint. Then I render these in my slow cooker set on low (you could also do stovetop over very low heat). When the skins are turning crunchy and all the fat is rendered out, I strain the schmaltz (as chicken fat is called in Jewish cooking) into a glass jar

91

and store it in the fridge. This works even better with duck than with chicken; and trust me, potatoes pan-roasted in duck fat are fantastic.

I mentioned in the first chapter that limited use of animal fats is not the health hazard many of us grew up believing it was. If you're skeptical, I highly recommend Sally Fallon's well-researched *Nourishing Traditions: The Cookbook That Challenges Politically Correct Nutrition and the Diet Dictocrats*.

When I was a kid, we used to save bacon fat, but I don't remember my mom or dad ever actually using it for anything. I think that for them it was just a reflex left-over from growing up with Depression-era parents who saved the drippings so that they wouldn't have to spend money on cooking oil. Well, a locavore in the Northeast has a reason beyond saving pennies to do so—there isn't any local cooking oil. As I mentioned before, there could be because oil-rich seeds and nuts can grow here, but nobody is doing anything with them commercially yet. There is fabulous local butter, but that's it as far as cooking fats unless you render your own.

Share a cow (or a goat). If you've got room in your home for an extra freezer, consider going in with a few friends on a half or whole cow or goat. I'm not joking. My friend Ellen does this and gets superb local, free-range meat at a significantly lower price than its equivalent from other sources. No, you won't have to hang half a beast somewhere in your home: The meat comes pre-butchered into chops, steaks, and other familiar cuts. Some CSAs offer this option, but an online search may turn up other farms near you that do.

Eat at Home (Takeout Doesn't Count)

If your lifestyle has included more than one takeout or delivery meal a week, frequent meals at restaurants even slightly more expensive than fast-food chains, or lunches grabbed at the corner store, then cooking at home will definitely save you money. If you consider yourself more of an assembler than a cook and haven't a clue what you'd eat at home if you had to prepare it yourself, please read chapter 8, Making Friends with Your Kitchen. Single locavores who think it is too much trouble or too depressing to cook for themselves will find encouragement and specific ways to make it fun and worthwhile in chapter 11, The Single Locavore.

You may start out eating at home to save money, but soon you'll find there are more benefits than that. If you share your home with a partner, family, or roommate, the time spent preparing the meal becomes as much of a social occasion as the actual sitting down to share the meal. You'll find yourself swapping stories about the farmers you've met who grew the food you're about to eat and brainstorming recipe ideas. In other words, eating at home is not just about saving money and keeping your body fueled. It is also good life.

IF YOU DO JUST ONE THING...
How to smart shop at a farmers' market

In addition to the suggestions above, the easiest and most helpful thing you can do to stick to a budget at a farmers' market is . . .

1. *Walk through the whole market before you buy anything. Prices for the same ingredient will vary from vendor to vendor. For example, yesterday at the Grand Army Plaza market in Brooklyn, Yukon Gold potatoes were $3 per pound from one farm but only $1.50 per pound at another.*

2. *Leave your grocery list at home. Instead, buy what looks the best and is reasonably priced, then plan your cooking around that. Why? Well, if you show up at the market having already decided that you're having green beans with dinner and green beans aren't really in peak season yet, they're either going to be frustratingly unavailable or priced high. Not sure how to invent a recipe based on what you brought home from the market? Just go online and do a search (e.g., "green bean recipes"). Or look up your ingredients in the index of* Joy of Cooking *(the one cookbook I'd save if I had to toss all the others).*

7
The Convenience Factor

This chapter is about how to handle hectically scheduled days and tired nights while still eating a local foods diet. This is the chapter I wish someone else had written before me so that I could have benefited from its advice. Figuring out how to deal with the convenience factor required more changes in my lifestyle than any other aspect of becoming a locavore.

"What is the hardest part?" is one of the first questions people ask when I tell them about being a locavore. My answer: planning ahead. For thirty years I've lived in "the city that never sleeps." When I had a nonstop runaround workday and didn't think to bring any food with me, well, lunchtime or midnight, there was always food just a few doors away. And if I got home too tired to cook, that's why they invented takeout and pizza delivery, right? All of that changed when I decided to go local.

The biggest temptation to cheat on my commitment to local food happens when I am hungry, tired, and/or short on time. It hits on those mornings when my alarm goes off at 6:30 a.m. and I hit "snooze" not once or twice but three times. I scramble out of bed with just enough time to get dressed and make it out the door. But I need breakfast, and there won't be any local food options at Adelphi University, where I'm scheduled to teach all day. Temptation also grabs me when I'm stuck waiting on a crowded subway platform for a delayed train and I can feel my stomach rumbling with hunger. The salted cashews at the platform bodega look really good, and I debate whether it's worth compromising just this once. During The 250, the answer was always no, it wasn't worth it. Since The 250 ended, I am less of a purist. Once in a while I do cave in and go for the convenience food. I'm not perfect. But I'm proud to say that doesn't happen often, because during The 250 I learned how to make my local foods diet convenient for just such "emergencies." It does take planning ahead, and as I said, that's not something I was used to.

Planning ahead doesn't just mean packing lunch and making sure there's something in the house for dinner. Because I live in a region where January temperatures can get down into the single digits, planning ahead includes thinking in summer about what I will eat in winter. That kind of foresight definitely required a learning curve. It's one thing to put up a few cute jars of jam to give as gifts, entirely another to know that if I don't can and dry some tomatoes in late summer, I won't be eating any tomatoes until the following July. And just how much food do I really need to preserve anyway? My generation didn't grow up knowing how many jars of home-canned tomatoes one person needs to get through the winter.

Planning ahead also includes knowing what I can get where and when. It means including on my iCal alerts that the farmer who sells wheat flour is only at the market on Friday, so I'd better get there on Friday if I want to have bread or pasta this coming week.

Before you decide this whole locavore thing sounds like too much trouble, let me assure you that there are ways to honor a commitment to local foods even when you don't have a minute to spare. Some of them require the planning ahead I've been going on about, but others can help you even when you didn't plan ahead. It took a lot of trial and error to figure out how to deal with the convenience factor, but it got easy for me eventually. The information in this chapter will make it easier for you too.

Easy is important sometimes. Humans tend to opt for convenience whenever it is an option. Until the Industrial Age, though, it wasn't an option for most folks. There wasn't restaurant food delivered in disposable containers or a twenty-four-hour deli. There weren't long aisles of frozen and canned food at the store. But starting with the Industrial Age, scientific innovations such as petroleum-powered vehicles and refrigerated train cars, plastics, and the whole concept of "disposable," led us into a new era. Now it wasn't just the rich and powerful who could opt for convenience. Hearing about my enthusiasm for local foods, one woman of my grandmother's generation said, "Well, good that you're doing that, but I'm really glad I don't have to slave in the kitchen like my mother did." FYI, I don't "slave in the kitchen," but I got her point.

I grew up in a time when convenience was king because ignorance was bliss. The first dinners I cooked for myself when I moved out on my own were Stouffer's frozen meals, heated up and served in the aluminum tray they came in. As for thinking about where the ingredients in that meal came from, that wasn't on my radar at all.

But now we're living in a crossroads time. On the one hand, we are increasingly aware of the impact of our lifestyles, including food choices, on the world around us. The ignorance is gone and with it the dubious bliss. On the other hand, many of us still reach for the prewashed, plastic-wrapped salad greens from across the country when we're in the produce aisle. We do know better now, but that alone is not always enough to motivate us to make a change. We're addicted to convenience. It's like telling smokers that their habit is killing them: Hearing the truth doesn't necessarily convince them to stop. Telling people that the way they eat and live is killing the planet doesn't necessarily get them to give up Starbucks in a paper cup, never mind the out-of-season food.

I understand. I am a similar bundle of contradictions, and I'm grateful for many conveniences. I'd rather turn a knob on my stove than have to build a fire every time I want to cook. I think it rocks to be able to run hot and cold tap water. I'm happy to drop off my recycling in the bins outside my apartment building rather than figure out how to transport them to the nearest recycling station. (Like many New Yorkers, I don't have a car.)

The first nonlocal food I ate after my 250-mile diet year ended was a choice made entirely for the sake of convenience. My dad was visiting and we'd thrown a party the night before to celebrate the successful completion of The 250. There were about a dozen of us in the garden, drinking local wines and relishing the blowout local foods feast we'd prepared. For the party we made several dips, as well as home-made sourdough crackers. The crackers disappeared early on, but there was still a little dip left over the next day. Dad and I wanted crackers and dip, but crackers take time to make and we were hungry. During The 250 I would have just said never mind to the cracker craving and instead had carrot sticks or radishes to go with the dip. But The 250 was over. Dad went out and bought a box of crackers—the first nonlocal food I'd eaten in a year. So I get the convenience factor, I really do.

Fitting the Farmers' Market into Your Schedule

Many city folk have gotten used to shopping only when they run out of something. This is a bad idea if you are a locavore. I found this out during The 250 when I ran out of onions. Almost every savory recipe seems to start with chopping an onion. My CSA share hadn't included onions that week, the nearest farmers' market wasn't for six more days, and I didn't have time that week to get to one of the markets that required a train trip. The lack of onions taught me to stock up on certain staples (see The Locavore's Pantry below) and not to skip my neighborhood market on Saturday.

Some people have the opposite shopping pattern. Instead of only hitting the store when they need something, they've gotten used to big, infrequent trips to the mega-mart. They go to Costco or Walmart and load their carts so that they won't have to shop again for weeks. That's why American refrigerators are so huge, by the way. They aren't in other countries, where people value freshness over convenience.

If you've got a big family, then shopping this way even at the farmers' markets may make sense, but for most people it doesn't. Fitting smaller, more frequent shopping trips into your schedule may seem like a time suck, but it doesn't have to be if you cut the transportation time to a minimum.

In the New York City area there are farmers' markets every day of the week in every borough. The key is to find out which ones are near you on which days. What you don't want to do is spend hours getting to and from a market that isn't near you just because you ran out of something.

The Council on the Environment NYC's Web site has a map of greenmarkets that you can print out. It shows where the markets are and which days of the week they are held. Look for markets near your workplace as well as the ones near your home. If there's one a few blocks from your office on Tuesday, then that's the day you bring some extra bags and stop by on your way to or from work. If it will be a long time before you can get the fresh produce or other foods home, invest in a few thermal bags that will keep your food cool for hours (see Useful Resources).

Think you might get poked fun at if you show up at work with a bunch of kale and a carton of eggs? Don't worry about it. Brush up on those talking points at the end of chapter 2, and start spreading the word.

What's in a Locavore's Bag? Stuff You'll Be Glad You Brought with You Every Day

A Snack

I've already mentioned my biggest temptation: those roasted, salted, imported, cellophane-wrapped cashews at the bodega on the subway platform. If I'm hungry and stuck waiting for a train, it's just too tempting. Which is not to say that I never treat myself to cashews. They are one of my favorite special treat foods, but the ones from the subway bodega aren't even very good—they're usually slightly rancid. So I hold out for that cashew treat until I can get some really good ones and take my time enjoying them rather than just shoveling them into my ravenous maw.

The trick is to always have a snack with me just in case. My favorite pack-in-my-bag snack to fend off a bodega attack is dried fruit (usually apples). They keep pretty much forever, so if I don't eat them one day, I don't even bother taking them out of my bag. I just leave them in there for another day.

Bags
The other thing I always have with me is cloth and plastic bags. I reuse the plastic bags until they are falling apart and then recycle them at the supermarket. (Did you know that supermarkets in New York are now required to have a plastic bag recycling bin? It's often hidden somewhere behind the cashiers, so ask if you don't see it.) Very few plastic bags come into my world anymore, so most of the time I carry reusable, washable muslin produce bags.

Bags are useful for unplanned trips to the food co-op or a farmers' market, and they save me from either having to run home to get bags or feeling guilty for taking new ones. I started making sure I always had bags with me because of an embarrassing video. Kitchen Caravan filmed me for their online cooking show, and the spot focused on the fact that I was a locavore. It included an interview in which I touted all the environmental benefits of eating local foods. The camera followed me to the farmers' market. They wanted to get shots of me buying ingredients for one of the recipes we'd be shooting later, a Greek potato and garlic dip called *skordalia*. So there I was reaching into my purse only to realize that I didn't have any bags with me that day. I grabbed one of the virgin plastic bags the farmer had near the potatoes and we went ahead with filming the segment. Boy, did I get a lot of flak from environmentally concerned cooks who saw that show, and rightly so.

A Notebook and Pen (or Your PDA)
I never know when I'm going to stumble upon a treasure like a store I've never been into before that carries Old Chatham's Ewe's Blue cheese, or a bakery that features bread made with locally grown grains. I don't trust my memory to keep track of all my finds, so I take notes.

A small notebook in my back pocket is how the Locavore's Guide to NYC started, although nowadays I'm more likely to jot my finds down in my iPhone. Use your favorite, handiest method for leaving yourself notes. It will save you time in the future when you're trying to remember where you saw that great local bread.

You can start your own customized version of the guide, or you can use (and contribute to) the online version at www.localfork.com.

The Locavore's Guide to NYC

The Locavore's Guide to NYC is an online directory of local foods organized by product. You can look up "feta cheese," for example, and quickly find out where to get it on what day of the week.

The guide started out of pure frustration. During the second month of my 250-mile-diet year, I made two really stupid trips to the Union Square market that were anything but convenient. The first time I had a movie night at home planned and wanted some popcorn. I'd spotted popcorn at the greenmarket, so I thought it would be simple to add a trip to the market as an extra stop on the way home after work. But I had noted neither the day I'd spotted that popcorn nor the farmer's stall that was selling it. I got there on a Saturday afternoon, and if you've ever been to the Union Square market on a Saturday, you know how huge and crazy-crowded it is. Fun if you're just there for a stroll and to see what's available, but I was on a mission. I wasted a lot of time walking through all the stalls. No popcorn. Finally I stopped at the information booth and found out that the farm I'd seen selling popcorn wasn't at the market on Saturday. Aargh! (By the way, you can now get popcorn at Union Square on Saturday. Oak Grove Plantation carries it.)

You'd think I would have learned my lesson, but the same thing happened when I ran out of flour and wanted to make pasta. No problem, I thought. I have a little extra time today, and I know there is flour at the greenmarket. But once again I couldn't remember which farm sold it or what day I'd seen it there. I went on the wrong day—and made the trip for nothing.

I started carrying a small notebook in my back pocket whenever I went to the farmers' markets. I jotted down what was available, which farms had certain items, which market I'd seen them at, and on what day of the week. I also took note of where each farm was located and, when I got home, looked up how far they were from New York City. My entries looked something like this:

Bread flour

Wild Hive Farm (70 miles)

Union Square market Wednesday and Friday, but only
April–June and September–November

My pocket notebook filled up quickly, and I thought the information I was compiling could be a real time-saver for other locavores in the area. At first I planned to publish it as a book, but I quickly realized that wouldn't be practical. What's available when at which market changes all the time, so a print version of the guide would rapidly become out of date. I mentioned my pocket guide at a Food Systems Network NYC meeting. A woman from Local Fork at the meeting thought she could turn it into an online directory. Within a few weeks we had it up and running.

The guide is constantly being updated as new local products become available. Since I can't scout every borough for local foods by myself, the online guide is set up with e-mail links you can use to let me know about your own local food discoveries. Please consider being a scout and letting me know what you've found. I'll use your information to update the guide, and with your help it will make the locavore life much easier for all of us.

In addition to using the guide online, you can start your own pocket version that targets the foods and farmers you like best. If you want to check how many miles a certain farm is from you, I recommend using the online City Distance Tool (see Useful Resources).

Prep Your Produce Now So You Don't Have to Later

When you get your local harvest home, there are ways to use and organize it that will make your life much easier. Washing and drying your produce before putting it away takes time, no question, so wouldn't it be more convenient to just shove it into the refrigerator and get to it later? Not for me. Let's say I get home late and tired. Which am I more likely to use, the already washed and ready to go salad greens or the head of lettuce that needs to be broken up, washed, and dried in my salad spinner? A few minutes up front prepping my produce definitely makes all the minutes later on easier.

When I come home from the farmers' market or have just picked up my weekly CSA share, I plan on spending a little time getting all of that bounty into ready-to-eat shape. It doesn't take a ton of time: About half an hour a week will do it. Although I may be tempted to shove the food into my crisper drawers and get to it later, when later comes, a container of clean carrot sticks is a handier snack that a bundle of roots with the dirt still clinging to them.

If you are getting your food from the garden, pick a little more than you need when you go outside to see what is ready to go into dinner tonight. Prep the extras and they'll be ready to use for lunch tomorrow, or for when it's pouring rain outside and you'd really rather harvest the contents of your fridge than the soggy outdoors.

Some vegetables and fruits keep better if they are not washed before storing. Here's what is worth prepping and what is not:

Prep Ahead

- **Salad greens.** Wash and spin dry.
- **Snap beans.** Wash and snap off the stem ends so that all you have to do is throw them in a steamer or bit of water for a quick side dish.
- **Crudités.** Wash or peel carrots, radishes, kohlrabi, and other crunchy vegetables that are good with dips or for munching as is.
- **Fruit.** Wash. Exceptions are berries, as well as any stone fruits such as peaches or apricots that you are keeping at room temp to ripen. These are better left unwashed until you are ready to eat them.
- **Herbs.** Wash. There's nothing like a little crunchy soil residue to go with your cilantro, right? Yuck. I'm much more inclined to add generous sprigs of fresh herbs to my food if I've either just picked them from the garden or they're already washed and ready to use. To store bunches of parsley, cilantro, basil, and other herbs, wash them, shake dry, and put in a cup with an inch of water. Cover with a plastic bag and store in the refrigerator. Another method is to loosely

wrap herbs in a damp paper or cloth towel and then store them in a sealed plastic bag in the refrigerator.

Don't Bother

- **Root vegetables,** except the ones you will be using for crudités. Beets, parsnips, carrots, etc., store best in the refrigerator unwashed. As mentioned above, I do wash and slice some of my carrots and radishes so they are ready to eat. All other root veggies I put away unwashed.
- **Braising greens,** or leafy greens you will eat cooked. I don't prewash these because in many recipes, the water that clings to them after washing is all the liquid you need to cook them in. There's no point washing them until just before I'm ready to cook them up. However, I do recommend that you cook leafy greens as soon after you bring them home as possible. Steam, sauté, or boil them until just tender then run under cold water to stop the cooking process. Squeeze out excess water (squeeze hard). Store in the refrigerator and they will last for a week, ready to add to soups, omelets, dips, and many other recipes.
- **Cabbage.** Cabbage stores longest whole, so don't bother washing or cutting it until you're going to use it.

Prioritize Your Produce

Put food that won't keep for long near the front of your refrigerator drawers, and tuck the stuff that is good for weeks farther back. Or allocate one drawer just to super-perishables such as salad greens and one for keepers like beets. (See chapter 8, Making Friends with Your Kitchen, for more suggestions on prioritizing your produce.)

Take It One Step at a Time (with Plenty of Time in Between)

Some food projects sound time intensive and exhausting if you think about doing them all at once. For example, in April my friend Ellen and I always forage for Japanese knotweed during its brief season. If I think about the whole project from harvesting to washing the knotweed to peeling it, chopping it, and then either freezing it or cooking it into a compote and then processing the jars of compote in a boiling water bath . . . sheesh, I'd need a whole day devoted to knotweed, and I don't have a day to devote to that.

103

Fortunately I don't need to. I meet Ellen in the park and we spend less than an hour collecting the knotweed, enjoying each other's company and catching up as we do so. Once home, I wash the knotweed and shove it into the refrigerator and get on with the rest of my day. A day or two later I get around to peeling and chopping it. If I'm making compote, I may not get around to it until the following day. And once I do cook it, the compote may go back into the fridge to be ignored for yet another day. Finally I reheat the compote, spoon it into canning jars, and process for a quick 10 minutes. Done. Yeah, that's a weeklong knotweed project, but I actually never spent more than a few minutes out of any day on it.

Many of my cooking and food preservation projects get spaced across several days in similar fashion. It is easy for me to find half an hour here, fifteen minutes there to spend on food, but it's often impossible for me to find a whole day or even several hours for such things. So the stock gets made overnight in the slow cooker, refrigerated, strained a day or two later, refrigerated again, and finally canned a day or two after that. Once I have the stock, the compote, or whatever else I've made, it becomes total convenience food: Just pop open the jar and use.

The point is, you don't have to do it all at once. It is always possible to sneak in a few minutes here or there for food projects.

With that said, I'm going to play devil's advocate for a moment: I don't know where we get the idea that preparing food is something unpleasant to get over with so that we can get back to the business of living. Who gains from convincing us that cooking is a tedious chore? The corporations who want to sell us their packaged convenience foods and foodlike products. The secret they don't want people to know is that preparing food can be a sensual pleasure, and fun.

When I get home with my CSA share, I know that I'll thank myself later if I get the prep done that evening. I don't see it as drudgery. I put some music on while I'm washing and peeling and bagging. If friends or family are coming over, I invite them early enough for us to chat while we put our meal together. I enjoy a little knife work, and chopping away at a carnival-bright pile of red, yellow, and orange peppers is a pleasure, as is the scent of the ripe peaches I just set on the table and the ruby glint of the stalks of that bunch of beet greens waiting to be washed. Why rush through as if the only point of food is eating it? The time spent on preparing food is not something to get out of the way so that I can get back to living my life. It is part of my life, an important and enjoyable part.

104

The Locavore's Pantry

I could have included this in the chapter on Making Friends with Your Kitchen, because it is about cooking staples. But the point of having these items on hand at all times is that there is always something in the house to eat. Inconvenience is realizing that I have to make a subway trip to a farmers' market to restock an ingredient I've run out of (or settle for a nonlocal version). With the things on this list, all I need is whatever fresh vegetable or fruit I've got on hand and I can come up with a meal:

- **Oil or butter or other cooking fat**. If you eat bacon or poultry, you can render their fat to make excellent cooking fats. (No, your arteries won't hate you—see Sally Fallon's *Nourishing Traditions* for why not.) After you cook bacon, pour the drippings through a paper or cloth coffee filter into a jar and store in the fridge. With poultry, snip off the fatty bits of skin and store them in the freezer until you have about two cups worth. (I know, sounds gross. But to me, wasting part of an animal that's been killed for my food is worse.) Render the fat over very low heat in a saucepan or in a slow cooker set on low. When the skin bits start to get crispy, it's done. Strain and store as for bacon fat above. The crispy bits are called cracklings and in past eras were considered a special treat. They still are.

- **Salt.** Although New York State used to be one of the leading producers of salt (from salt mines in the Finger Lakes area), it now only produces road salt for deicing wintry streets. The nearest source of salt I've been able to find is the Maine Sea Salt Company, about 450 miles from where I live in Brooklyn.

- **Onions** (or leeks or shallots, depending on what is in season and available)

- **Garlic** (or green garlic in spring or garlic scapes in early summer)

- **Honey**

- **Yogurt**

- **Eggs**

- **Beans,** especially precooked beans. Having some precooked beans in the freezer is almost as convenient as opening a can. You can give dry beans an overnight soak and then drain and freeze them so that they are ready to cook. Or you can go ahead and cook your beans after soaking and then freeze them. Frozen cooked beans do not need to be thawed before adding to recipes. Just dump the frozen clump of beans into a pot with a little water and simmer until warmed through. They are then ready to use for hummus-style dips, to add to soups, for Mexican-style beans and greens, or a hundred other uses.

- **Stock** (meat, poultry, fish, or vegetable. See chapter 8, Making Friends with Your Kitchen, and "Something for Nothing" in chapter 6, The Cost Factor). Stock can be frozen or pressure-canned for long-term storage. (*Note:* Stock cannot be safely canned in a boiling water bath.)

- **Salad dressing.** If you're used to keeping the bottled stuff on hand, having some ready-to-go salad dressing will make all those local greens more tempting to eat. Salad dressings are super easy to make. My house dressing goes like this: Whisk 1 teaspoon prepared mustard and 1 teaspoon honey and a pinch of salt into ¼ cup wine or cider vinegar. Add ¾ cup oil one splash at a time, whisking until it is completely blended before adding the next splash of oil. (Yeah, I know the cookbooks and TV chefs say to add the oil a drop at a time. Be my guest. I have other things to do.) Pour into a glass bottle. The mustard acts as an emulsifier and usually keeps the dressing blended, but if yours separates just give the bottle a shake before using. Keeps at room temperature for two weeks, in the fridge for up to two months. If storing in the fridge, you'll want to take it out half an hour before using so the oil can uncongeal. To vary the dressing, pour a little bit out and add herbs, mayonnaise, yogurt, or buttermilk. Don't add these to the bottle, or the dressing won't keep.

The Scrap Bowl

Cooking with fresh foods generates a lot of scraps: heel ends of root vegetables, eggshells, etc. Some of them can be used to make soup stock (see page 122), but many of them are destined for the compost. A scrap bowl makes it easy to deal with them.

I keep a covered pail in my kitchen for collecting compost scraps. It has a filter that keeps smelliness at bay, and it is big enough that I only have to carry it out to the compost bin in the garden every few days. But if I had to bend over and take off the lid and scrape stuff off my cutting board every time I needed to clear scraps to make room for the next item to be chopped, it would be tedious. Instead I collect them in a scrap bowl. This is just a medium-size bowl that I keep near my cutting board and work area. The scraps go in until it is full, and then it gets dumped into the compost pail, which eventually gets carried out to the compost bin.

If you're not composting, the scrap bowl will save you from multiple trips to the trashcan.

Cook (or Assemble) Ahead

When I do find myself with a little extra time and energy, I often cook something that's not meant for immediate consumption but will make life easier in the busy week ahead. Even if you don't cook much, you can still assemble in advance and ease any looming schedule crunch. Here are some suggestions:

Dips are especially easy because all you have to do is combine the ingredients, no cooking required. Once you have a dip or two in the refrigerator, all you need is some of those carrot sticks or radishes you prepped when you brought them home. (See chapter 8, Making Friends with Your Kitchen, for a simple dip recipe.)

A pot of soup can get me through a week of suppers, and it doesn't have to taste like the same leftover soup every day. My dad has a great phrase: "Expand the soup." To expand your soup, you just add something to it. Maybe it was straight up tomato soup the first day. The next day you add some greens and crumble on a little cheese. The day after that, a dash of curry powder changes the flavor of the soup completely. Chapter 8 has a great basic vegetable soup recipe with seven variations.

If you eat meat, cook an extra portion when you prepare it. The leftovers are great on salads or in sandwiches.

Speaking of Leftovers . . .

Your freezer is your friend. Leftover soup, stew, casseroles including lasagna, taco fillings, and other main courses can go straight into the freezer to be taken out for a no-hassle dinner weeks or months from now. (See chapter 9, Simple Food Preservation, for more ways to get your freezer working for you.)

During winter my freezer and shelves are packed with food that I've put up during the warm months. Truly, my freezer is dangerous in January. It isn't very big, and when you open its door there is every chance that you will be ducking under an avalanche of frozen blueberries and last spring's rhubarb.

I needed a way to keep track of what I had so that it would be easy to find whatever I was looking for, whether it was a jar of tomatoes behind a jar of green beans in the cabinet or the pork chop that's tucked behind all that frozen fruit. The Fridge, Shelf, and Freezer list is my solution.

Under a magnet on my refrigerator door is a piece of paper with several typed lists on it. There is a column for what's in the freezer; another for what's on my shelves; and during the CSA months, another list for what's in my crisper drawers. As weeks go by, I cross off things as I finish them, write in what I add. Eventually, when the page is so scrawled upon that it's almost illegible, I update the list on my computer and print out a new one.

A section of my list might look something like this:

Blueberries ~~2~~ 1

Rhubarb 1

Kale soup 1

Cooked black beans ~~3~~ 2

Chicken breasts ~~2~~ 1

The first time you make such a list is a bit of a pain—you have to take everything out of the freezer and off the shelves and write it down. But once you've got the list going, all you have to do is update it from time to time. Besides helping me find the ingredients I've got, the list prevents me from buying something I thought I'd run out of but actually still have plenty of.

Swapping for Shopping (and Maybe the Cooking Too)

If you really, really don't have time to get to the farmers' market or you really, really don't cook, there are ways around that. My friend Bill lives and works at Shen Tao Studio, an exercise and training studio just a few blocks from the fabulous Union Square Greenmarket in Manhattan. Four days a week, year-round, he has access to the largest farmers' market in the New York City area, with a wealth of culinary variety that can't be found at the supermarket. There are literally thousands of varieties of fruits and vegetables, including over one hundred varieties each of apples and tomatoes.

Unfortunately, Bill doesn't cook much. He wanted to take advantage of his proximity to the market but didn't want to spend time shopping or cooking. His solution is a barter agreement with one of his students. Every few days the student shops for local foods at Union Square, cooks in Bill's kitchen, and leaves him with a week's worth of food. In exchange, the student gets classes at Shen Tao for free. If you know someone with time and cooking skills, and you have neither, you may be able to trade whatever your expertise is for theirs.

Would You Eat This If You Had to Make It from Scratch?

In a radio interview, local foods guru Michael Pollan admitted that he has a weakness for potato chips. But he doesn't eat them often because of a personal food rule he follows. The rule is that if a food is such trouble to make that he wouldn't do it himself often, if ever, then that food is classified as a special occasion treat that he'll enjoy only rarely.

I like this rule a lot. It puts a whole other spin on "convenience." When I listened to that interview, it made me think of how many of our "convenience foods" are really not at all convenient to make. They are the result of considerable labor. Think of hamburger buns and Pollan's favorite, potato chips. How often would I really want to bake those buns, never mind grow and thresh and grind the wheat that turned into the flour that turned into the hamburger buns? Or those chips? I've made potato chips at home. They were pretty good, but they were also a lot of work and made a mess of my kitchen. Definitely not something I'd do every day.

"Do I want this badly enough to make it myself?" is a question I now ask myself when I'm staring at a box of crackers or some other convenience food in the store. Usually the answer is no. That doesn't rule out buying the inconvenient to make "convenience food," but it does remind me not to do it too often.

If You Do Just One Thing...

The Top Ten Grab-It-and-Go Local Food Snacks

1. *A piece of fresh fruit.*

2. *A bag of dried fruit (See chapter 9, Simple Food Preservation, for tips on drying the beautiful local fruit you bring home.)*

3. *A piece of cheese—maybe to go with that fruit? An apple and some local cheddar is one of my favorite winter snacks. Some pickled cherries with Old Chatham's Hudson Valley Camembert are heavenly too.*

4. *Prepped veggies. In addition to carrots and radishes, baby turnips, celery, peeled kohlrabi, and celeriac (celery root) are also good raw.*

5. *Dip (to go with those veggies).*

6. *Ratatouille, pickled green beans, or other home-canned foods that are ready to eat right out of the jar.*

7. *Yogurt with a couple spoonfuls of fruit preserves.*

8. *Any pickled vegetable—carrots, cukes, green beans—quickly mixed with some kernels of corn from the freezer for an instant corn salad. Add a little mayonnaise if you wish, or some beans if you have some already cooked.*

9. *Hard-boiled eggs. I make several at a time so they're always handy in the fridge to enjoy with a sprinkle of salt or to add to salads. One thing about fresh, local eggs, though, is that it's actually better to wait until the eggs are at least two weeks old before hard-boiling them. Older eggs are easier to peel. The ones at the supermarket are all relatively on the old side, so you don't really notice the difference. But if you boil some fresh CSA or farmers' market eggs, they really are problematic to peel (terrific for everything else though).*

10. *. . . and of course, leftovers can always be thrown into a container to go.*

8
Making Friends with Your Kitchen

When you eat local foods you will have adventures in the kitchen. Whether you're an experienced cook or someone who considers "I don't cook" an identity statement rather than a description of your cooking skills, read on.

Are there differences between a locavore's kitchen and most other contemporary kitchens? Yes, several.

One of the big differences is a relative absence of packaging and labels with long ingredient lists. This is great from an environmental perspective (less waste for the landfill), but it also means learning how to store fresh, unpackaged food so that it keeps until you are ready to eat it. Remember that local produce is picked at its peak. In the case of leafy greens, this means that properly stored they will keep longer than their supermarket cousins. But in the case of fruit, it can mean that you may have only a day or two to eat or preserve it before it starts to spoil.

Following recipes slavishly doesn't always work in the locavore's kitchen. Some ingredients behave a little differently from the standardized commercial brands. Local grains and flours can require less liquid than recipes call for. Recently harvested dry beans need shorter soaking and cooking times than most cookbooks specify.

A locavore cooks with what is seasonally available, which may not be what is written in the cookbook. That goes for the recipes in this book too. A recipe may call for a clove of garlic, but if I'm making the recipe in June when the farmers have run out of the garlic stored from last year's crop, and the new harvest won't be dug up until July, what do I do? Since it's June, there are garlic scapes available that can be substituted for "regular" garlic. Earlier in the year I might use green garlic or wild field garlic. In winter I dive into my pantry of preserved foods to replace some fresh ingredients, using home-canned tomatoes instead of fresh and dried celery in my soup stocks, for example. A little improvisation and creativity is a must in the locavore's kitchen, as is a willingness to substitute one ingredient for another.

Useful Substitutions for the Northeastern Locavore

There are certain ingredients that can only be grown in milder climates than ours. Citrus, for example, can't be grown here, so I try to keep my use of it limited to the special-occasions-only category. For everyday cooking I've come up with some alternatives to nonlocal ingredients that work really well. They don't taste exactly like the thing they are replacing, but they fulfill the same function in a recipe and are tasty in their own right. Most of them rely on my pantry of preserved foods. You'll find the recipes in chapter 9, Simple Food Preservation.

For lemons I use lemon-scented herbs for citrusy flavor. Lemon thyme and lemon balm are two of my favorites. I use vinegar for sourness, and I make herbal vinegar infused with lemon-scented herbs to combine those two qualities.

Instead of sugar, I use local honey. Honey is sweeter than sugar, so I use 25 to 30 percent less in recipes. I choose a mild wildflower or clover honey unless I want the taste of the honey to be the star in a particular recipe. Maple syrup is another local sweetener I love. It is also my favorite right-back-at-ya answer to West Coasters who brag about how they can grow everything there—they don't have maple groves and they don't produce maple syrup! But the distinctive taste of maple is so strong that I use it more as a flavoring than for its sweetness.

Capers are the immature flower buds of a Mediterranean vine. Any sharply tangy pickle, such as cornichons, can be finely chopped and used to replace capers in a recipe. You can also pickle green, immature nasturtium buds to make your own "capers."

I use chopped pickles to replace olives as well. This doesn't work well in salads, where the taste of an olive is too specific to replace, but is fine in pasta sauces and some North African and Middle Eastern–style recipes. The pickles add a similar salty tanginess.

Tomato paste could be made with local tomatoes, but it requires so many tomatoes and so much time to make that I don't bother. Instead I rehydrate dried tomatoes and mince them finely. The flavor is richer than the tomato paste you buy in a tube or can, and I've gotten to like it so much that I can't imagine why I would go back to the paste.

Cook's tip: The latest edition of the *Joy of Cooking* has a chart in the back with useful substitutions for dozens of ingredients.

Organize Your Produce

It's sad to get to the day before the next CSA delivery and realize that there is still a bag of lettuce in the back of the refrigerator drawer that is starting to spoil. Storing fruits and vegetables correctly keeps them fresh longer, preserves flavors and textures, and makes it a whole lot easier to make use of all the bounty you brought home from the market or your CSA distribution.

"Hard" vegetables—roots, winter squash, and some summer vegetables, including cucumbers, summer squash, and eggplant—keep longer than fruits and leafy greens. Leafy greens that are still in a head keep longer than the loose leaves in a salad or braising mix. So eat those loose-leaf mixes first, then the heads of lettuce, kale, etc., then the "hard" vegetables. Of course, you can mix them up as you go: no reason not to enjoy some of those beets on top of your salad as long as you remember that the salad greens have use-up priority.

Some foods should never be stored in the refrigerator because doing so changes their texture and flavor in bad ways. Potatoes should be stored at room temperature (or cooler, if you're lucky enough to have a cellar or unheated garage). They need to "breathe," so don't keep them in plastic bags. Potatoes start to sprout if exposed to light, so keep them in a drawer or a cloth or paper bag. Refrigerated strawberries and tomatoes are disappointing. Keep them at room temperature.

Other foods could be refrigerated but don't need to be. Onions, garlic, and shallots can all be kept out at room temperature. So can winter squash; but if you're going to keep it for longer than a couple of weeks, rub a little oil on the outside with a towel to prevent mold.

You Still Have to Wash It, Even If It's Organic

Those beautiful organic vegetables you just brought into the kitchen still need to be washed, even if they are just coming in from your garden. Remember soil and dust? Insects? Unless you like a little grit with your spinach or extra protein with your broccoli, give them a rinse.

What the Heck Do I Do with a Kohlrabi (and Other Unfamiliar Local Foods)?

A willingness to experiment and try new things is handy when you're swapping one recipe ingredient for another. It's also essential when you're dealing with an ingredient you have never seen, tasted, or heard of before. Conversations at our CSA distribution often include the question, "What is that and what do you do with it?" Faced with a garlic scape, a parsley root, or a kohlrabi, even experienced cooks may scratch their heads.

Getting to try ingredients you can't get at the average supermarket is one of the perks of local foods, but it is also one of its challenges. A search online will turn up plenty of options on how to use some of the quirky-but-tasty fruits and vegetables you may encounter. Here are a few suggestions to get you started:

Early to Mid-Spring

- **Fiddleheads:** These are the unopened fronds of the ostrich fern (*Matteuccia struthiopteris*). They must be cooked to be edible (Raw foodies take note—this is not just a matter of taste. Fiddleheads are slightly toxic until cooked.) Fiddleheads are delicious steamed or lightly boiled, and they make terrific pickles.

- **Green garlic:** This is simply young garlic dug up before the bulbs have fully formed. Green garlic has a milder flavor than mature garlic. In addition to the whitish underground part, you can use as much of the green part as is tender enough to slice easily. The tougher leaves can be used to make soup stock.

- **Ramps:** Also known as wild leeks (*Allium tricoccum*), ramps are in season for only a few weeks in early spring. Once you cut off the threadlike roots at the base, the entire remainder of the plant is edible, including the bulbs, which look something like scallions, and the leaves. Ramps have a distinctive yet gentle taste and should not be paired with strongly flavored sauces. You want them to be the main attraction of whatever you're cooking. A ramps omelet is heavenly, as are potatoes and ramps.

Late Spring to Early Summer

- **Baby turnips:** The difference between baby turnips and the big ones you get in the fall is that they are good raw. Usually they do not even need to be peeled. Just slice and add to salads or use instead of crackers with dips. Mature turnips are good in soups, mashed 50–50 with potatoes, or roasted with other root vegetables.

- **Currants:** These jewel-like fruits look like miniature grapes but have a tart rather than sweet flavor. Raw, they make a lovely garnish for desserts, but you probably wouldn't want to eat a bowlful of them. Cooked, they make excellent jellies and jams. They are high in pectin, and can help gel preserves made with low-pectin fruit such as strawberries. They also make excellent sauces and chutneys to go with lamb, pork, turkey, or game meats.

- **Fava beans:** Fava beans are labor intensive but worth the effort. Just don't plan on making them for a crowd. Like all shell beans (see below), they need to be removed from their pods. Unlike shell beans, favas also have a thick skin surrounding each individual bean that needs to be removed before eating. Cook shelled but unpeeled fava beans in a little water for 3 to 5 minutes. You can serve as is, peeling and eating each fat, tender bean as you go, or you can peel all the beans and serve them ready to eat. They are fantastic with a little grated cheese.

- **Garlic scapes:** These crazy looking curlicues with the pointy tips are the flower stalk and unopened flower buds of garlic plants. Farmers and gardeners cut them off when they appear so that the plants concentrate their energy on making bulbs (the "regular" garlic we're used to eating). The whole garlic scape is edible and has a strong garlic flavor, but without the burn that the underground parts of the plant have. Garlic scapes are best finely chopped and cooked in any way you would use garlic, though they can also be eaten raw.

- **Kohlrabi:** Many people think that these green or purple orbs are roots, but actually they are one of the aboveground parts of the plant. Kohlrabi is in the same plant family as broccoli (the Brassicaceae) and has a similar flavor when cooked. Raw, it is pleasantly crunchy and tastes a little like a mild radish. I usually peel kohlrabi with a paring knife rather than a vegetable peeler, because the outer eighth of an inch is fibrous and tough. However, one fellow CSA member tells me that unpeeled kohlrabi is good when she runs it through the shredder blade of her food processor. She then uses the shreds as a salad topping.

- **Lamb-quarter:** This is a common weed with a mild flavor similar to spinach. Lately I've seen it at food co-ops and a few farmers' markets. Maybe the farmers are weeding out their rows and selling us the weeds—and good for them, because lamb-quarter is delicious. Use the leaves and tender stems, discarding any tough stems. Like spinach, lamb-quarter cooks down to a fraction of its raw volume, so start with at least five times as much as you want to end up with.

Late Summer to Early Fall

- **Elderberries:** Often found growing wild, elderberries sometimes make an appearance in CSA shares or at farmers' markets in late July and August. They are a bit bland when eaten raw, but are wonderful in pies, jellies, and homemade wines.
- **Shell beans:** These are just young bean pods with fully formed beans inside. Left on the vine, they mature into dry beans. They have a more tender texture than dry beans—and a much shorter cooking time. There is no need to soak them before cooking. Simmer shell beans (out of their pods, or "shelled") in water for 20 to 30 minutes.

Fall and Winter

- **Celeriac (celery root):** This gnarly looking root has a thick outer layer that needs to be removed with a paring knife. Celeriac has a strong celery aroma that becomes milder as it is cooked. Roast celeriac with other root vegetables, add to soups, or cut into thin strips and toss, raw, with salad dressing or remoulade sauce.
- **Gobo:** This is the root of the burdock plant, and is a well-loved vegetable in Japan. Leaving it unpeeled gives you a vegetable with a musty, almost mushroomy flavor. Peeling it makes for a milder taste. Thinly slice gobo and add to stir-fries.
- **Parsley root:** Looking like a tangle of thin parsnips, parsley root is the forgotten edible part of the parsley plant. Chop it up and add it to soups and stews.
- **Rutabaga:** No, that's not a monstrously oversized turnip, it's a rutabaga. Peeled and chopped, rutabagas can be used in all the ways mature fall turnips can be. They are also good mixed with shredded apple and baked or fried.

I'm an Assembler, Not a Cook

Several of my friends consider themselves "assemblers" rather than cooks. If this sounds like you, read on for some simple ways to put together your local foods meals.

In some ways, working with local foods is easier on the new cook because the ingredients are so fresh and flavorful. My cooking is simpler now than it was before I became a local foods enthusiast because the tastes are so vibrant, I don't need to drown them in fancy sauces.

Summer is the assembler's season. Perfectly ripe tomatoes don't need much beyond a few torn basil leaves and some local mozzarella or feta. A ripe peach needs nothing at all. The ingredients of spring, fall, and winter require a little more time but can still be prepared simply.

Combine local ingredients to create dishes as enticing as any cooked menu. Try local cheese with seasonal fruit, salads topped with the ready-to-eat local smoked trout, tomatoes and cucumbers with fresh herbs, or strawberries briefly soaked in local honey and wine.

The dip below is perfect for assemblers. It can be served with different veggies depending on the season. Peel and slice kohlrabi, carrots, baby turnips, cucumbers, or radishes and use them instead of crackers. Raw snow or snap peas are delicious with this dip. I've suggested a few variations that just require a sprig of herbs or a pinch of spices to dramatically vary the taste.

Yogurt Dip to Go with Whatever You've Got

MAKES APPROXIMATELY ½ CUP. RECIPE CAN BE DOUBLED.

Place a strainer or cone-shaped drip coffee filter over a cup or bowl and line it with a paper or reusable cloth coffee filter. Put in a pint of local yogurt. Place the whole setup into the refrigerator and leave it there for twenty-four hours.

The strained yogurt will have a consistency like soft cream cheese. Scoop it out into a bowl. Add ¼ teaspoon sea salt and either ¼ teaspoon garlic powder or 1 clove of garlic very finely minced. Stir, cover, and let sit in the refrigerator for at least 2 hours before serving for the flavor to develop.

Variations: You can add approximately 2 teaspoons of any fresh herb, minced, to the dip. Chives, cilantro, parsley, or mint are particularly good. Don't drive yourself crazy measuring the herbs; just eyeball it. A little more

> or less won't hurt. Another variation, though it uses nonlocal spices, is to add ¼ to ½ teaspoon curry powder. If you like spicy food, try adding a pinch of ground cayenne pepper.

Trust Your (Common) Senses

If it smells like it's burning, it probably is burning, and never mind that the recipe specified 10 minutes over high heat.

One of the things about cooking is that however scientific and precise modern recipes try to make it sound, there are still a lot of subjective choices to be made. How high is "high heat"? It depends on your stove. How big is one "medium" onion? What if my clove of garlic is bigger than the one the recipe writer used? What does "until lightly caramelized" look like? These are things that you learn by experience and by trusting your senses. All good cooks have stories of culinary mistakes. That's how they learned to be good cooks. So if you're just beginning to cook, cut yourself some slack.

Kitchen Essentials

Let's start with the one cookbook I would have if I could have only one, *The Joy of Cooking*. There are very few recipes that you can't find in this book, and most important it gives clear explanations of how to handle different ingredients and why certain cooking methods do what they do. If you don't already own a copy, get one.

As for cooking equipment, there are some things on the list below that might not be considered essential in a "regular" kitchen, but they are huge time and effort savers for locavores (the salad spinner, for example). I'm sure other cooks would add their favorites to this list, but these are the things I use almost every day:

Must-Have

- Chef's knife—a big 8- to 10-inch knife. If you're going to spend a little extra, a good chef's knife is worth the investment. This will be your go-to knife for almost everything.

- Paring knife. Some jobs are easier with a small paring knife.
- Large, heavy pot such as a Dutch oven. This is the other thing besides the chef's knife that I consider worth spending extra on. A good, heavy pot such as an enameled cast-iron pot distributes heat more evenly than cheap, light pots, which means food is less likely to burn.
- Smaller pot or saucepan
- Frying pan, nonstick. Regular frying pans are better for browning meats and vegetables, and it's nice to have both; but if I had to choose one it would be nonstick.
- Cutting board(s)
- Kitchen towels (can double as potholders)
- Colander
- Wooden spoon(s)
- Spatula(s)
- Whisk(s)
- Ladle
- Measuring cups and spoons
- Grater
- Baking dish or tray for roasting vegetables
- Glass jars for storing, preserving, and freezing foods

Good to Have

- Pastry scraper. No, not for pastry. This is by far the easiest way to convey chopped food from your cutting board to the pan on the stove.
- Freezer bags or containers
- Strainer
- Food processor
- Immersion blender
- Salad spinner
- Timer

Serving Suggestions for Each Season's Quintessential Produce

Spring: Leafy Greens

When the first CSA distributions of spring arrive, they are green. Very green. Lettuce, kale, chard, bok choy, mustard and dandelion greens . . . I find that I have to eat leafy greens every single day in order to keep up (a nutritionist's dream).

One of the most satisfying recipes for leafy greens is also one of the easiest: Sauté minced garlic in a little oil or butter. Add washed, torn, or chopped greens and cook, stirring, until wilted. Add a little water if the greens start to stick. Season with salt to taste and a sprinkle of vinegar.

As I mentioned in chapter 7, often I steam or boil greens until just wilted and tender. I run these under cold water to stop the cooking and then squeeze out the excess water. These cooked, unseasoned greens keep in the refrigerator for up to a week and are ready to add to frittatas, soups, quiches, dips, and other recipes. If the week is nearing its end (and the next installment of CSA greens is on its way), the cooked greens can go straight into the freezer without further blanching.

Summer: Fruits, Including Tomatoes, Etc.

Summer is fruit season, which according to the scientific definition includes any seed-bearing part of a plant. The terms "fruit" and "vegetable," used to mean sweet versus savory, are cook's terms, not botanical ones.

While there are thousands of recipes for summer fruits (including the ones we think of as vegetables), I'm only going to give you two suggestions here:

The best thing to do with ripe summer fruits, including tomatoes, is to eat them as is. Sure, I sometimes use them in recipes just for the fun of it and for variety, but the only essential thing to know about these foods is how to choose the best, juiciest, perfectly ripe ones. A tip that will help you do that is to use your nose. Supermarket fruit is bred for uniformity and appearance, but a cantaloupe should smell like a cantaloupe and a tomato should smell like a tomato. When it comes to predicting flavor, smell is often more accurate than vision.

Other summer fruits are deliberately picked and meant to be eaten while they are underripe. These are the ones we think of as vegetables and include eggplant, squash, and snap beans. With the exception of eggplant, almost all summer fruits-known-as-vegetables need is to be steamed or boiled for a few minutes and then tossed with a little salt, butter, and maybe a light sprinkle of fresh herbs. Eggplant needs to be roasted or pan-fried until cooked through and then can be used in ratatouille, dip, and casseroles.

Fall: Root Vegetables and Winter Squashes

All the root vegetables and winter squashes (butternut, acorn, pumpkin, etc.) are excellent roasted. Peel and chop them into 1-inch chunks. Toss them in a baking dish with a little oil (rosemary oil is especially good), salt, and a clove or two of peeled garlic. Roast in a 400°F oven for 30 to 45 minutes until the vegetables have some browned spots and are soft when pierced with the tip of a knife.

Winter

"What do you eat in winter?" is a question I hear often. December through March is the season when the kitchen of a locavore living in a cold-winter climate least resembles today's conventional kitchens. Although all the animal products, including dairy and eggs, are still available, there isn't much fresh produce: a few fresh greens from local farmers who have unheated greenhouses, stored foods such as root vegetables and apples, but not much else.

The staples of my kitchen in winter are the same as they are the rest of the year: flour, grains, beans, onions, garlic, etc. What changes is that my store of preserved foods comes into its own at this time of year. Instead of tomato salad, I have tomato soup made with home-canned tomatoes. Instead of a fresh peach for breakfast, I make a smoothie with yogurt, honey, and peaches I froze last summer. I perk up potato salad with corn relish, meats with chutney. Instead of lettuce salads, I have chopped apples with sauerkraut, or dilly beans straight out of the jar. I make soup stock with dehydrated celery and carrots instead of the fresh stuff.

It all tastes good, but different. I don't expect canned tomatoes to taste like fresh tomatoes any more than I expect a raisin to taste like a grape.

(See chapter 9, Simple Food Preservation, for more winter pantry ideas.)

Vegetable, Poultry, Meat, and Soup Stocks

Homemade stock is very easy to make. Yes it takes a long time, but you don't need to be hovering over the stove during the whole process. I usually make stock in my slow cooker overnight while I'm sleeping. Sometimes that lazy method backfires, because the smell of cooking chicken stock drives my cat crazy. She won't let me sleep a wink until I get up and give her a midnight snack.

Basic ingredients for any kind of stock
Celery
Onion (If you save scraps of onions, carrots, and the green parts of leeks and other alliums as described below, you will rarely need to use a virgin onion for stock.)
Carrot

Optional ingredients
Herbs, including bay leaves (northern bayberry if you want the local option)
Tomatoes
Leeks
Poultry, meat, or fish bones (and fish heads)

1. Stockpile your vegetables and bones (if using) in clearly labeled containers in the freezer. Include the stems of herbs saved after you've used their leaves in other recipes; green parts of leeks and green garlic; heel ends of onions, carrots, and tomatoes; fennel fronds and celery leaves. Make sure all the vegetables and herbs are clean before they go into the freezer.

2. Bones can be leftover from cooked meat or raw. If you buy whole chickens and cut them up yourself (much cheaper than buying parts), you can use the backs for stock. Store poultry, meat, and fish bones separately, adding vegetable and herb scraps as you have them. I don't eat beef often

enough to get a lot of beef bones, so I combine beef, pork, and lamb bones. When making stock with bones, add a splash of vinegar while it's cooking. You won't taste the vinegar in the finished product, but it helps release calcium from the bones, resulting in a more nutrient-rich stock.

3. Place your ingredients in a large pot or slow cooker. Add enough water to cover. If making your stock on the stove, bring it to a slow simmer. Don't let it come to a boil or you will have a cloudy stock. If using a slow cooker, use the low setting.

4. Cook for 8 hours (give or take a couple—stock is forgiving) for meat and poultry stocks, 2 to 4 hours for vegetable and fish stocks. Strain through a fine mesh strainer.

5. Stock can be frozen for future use or will keep in the refrigerator for one week. To store at room temperature, stock (including vegetable stock) must be processed in a pressure canner. A boiling water bath is insufficient to safely preserve stock. Pressure-can pints of stock at 10 pounds of pressure for 20 minutes.

Asian-Style Nasty Bits

In chapter 6, The Cost Factor, I suggested that meat eaters save money by cooking up some of the offal, aka organ meats. Here is a recipe for those "nasty bits" that you and your guests will gobble up (your call whether to tell them what it's made with).

SERVES APPROXIMATELY SIX

½ pound livers, hearts, and/or kidneys
½ teaspoon local honey
1 tablespoon dark soy sauce (*Note:* No local version is available, for those of you who are being strict about this.)
1 tablespoon toasted sesame oil (ditto)
2 cloves minced garlic
1½ teaspoons vinegar
1 tablespoon water
1 handful fresh cilantro, finely chopped

1. Place a medium-size pot of water on the stove and bring to a boil. Meanwhile, chop the meat into approximately 1-inch pieces (no larger), discarding any sinewy or tough parts as you go. When the water is boiling, add the chopped meat and cook for 2 minutes.

2. While the meat is cooking, combine the rest of the ingredients. Drain the meat in a colander, then mix well with the sauce. Let sit for at least 10 minutes before serving with toothpicks or forks.

Can be made ahead and served at room temperature the following day, assuming you don't eat all of it for a snack before the guests arrive—it's that good.

Basic Vegetable Soup

MAKES APPROXIMATELY 7 CUPS (2½ PINTS)

2 tablespoons olive oil or butter

½ cup diced carrots (about 1 large carrot)

1 cup diced onion (about 1 medium large onion)

1 cup diced celery (2–3 stalks)

1 clove garlic, peeled and minced

¼ cup dry white or red wine (optional)

4 cups hot stock or water

2 cups chopped tomatoes or 1 pint canned tomatoes

1 cup diced potatoes or 1 cup diced turnips or radishes

1 bay leaf (northern bayberry, if you want to use our local variation)

1 teaspoon salt

2 tablespoons chopped parsley

Salt and pepper to taste

1. Heat olive oil or butter in a large pot. Add carrots, onions, and celery; cook on medium heat until they begin to soften, stirring occasionally.

2. Add garlic and wine and cook, stirring, for 1 minute.

3. Add stock or water, tomatoes, potatoes (or turnips or radishes), bay leaf, and salt. Increase heat to high and bring to a boil. Reduce heat to low; cover and cook about 30 minutes.

4. Add parsley, salt, and pepper.

Novice Cook's Tip: Don't worry about how small the pieces need to be to equal "diced" in this recipe. Whatever size you would like to find on your spoon when you sit down to eat the soup is the right size. Larger pieces will require a little more cooking time. Taste them, and if they are soft enough to please you, they're done.

Velvet Soup

For each pint of Basic Vegetable Soup, peel and chop the following into approximately ½-inch chunks:

1 medium sweet potato or 1 cup diced winter squash or 1 cup parsnips

1. Steam or boil until very soft. Add to Basic Vegetable Soup, above. Puree in the same pot using an immersion blender or process in a blender until pureed. Add salt if needed.

2. Serve with a sprinkling of fresh or dried thyme leaves or parsley and (optional) a dollop of yogurt or sour cream.

Curried Velvet Soup

Prepare Velvet Soup, above. Add:

½ teaspoon curry powder (Okay, the spices in curry powder aren't local. Purists, move on.)

Serve with a dollop of yogurt or sour cream and a few sprigs of fresh cilantro.

Bean Soup with Pesto

Prepare Basic Vegetable Soup, above. Add:

2 cups cooked beans
Pesto or extra-virgin olive oil and basil

To serve, add a teaspoon of pesto to each bowl of soup. If you don't have time to make pesto, drizzle on some extra-virgin olive oil and tear up some basil leaves to sprinkle into each bowl. (Yeah, I know the olive oil isn't local. It's one of my personal exemptions, remember?) Grate a little cheese over the top. (Sprout Creek Farm's Barat and Rockhill Farm's Equinox cheeses are fantastic local options, but you could use Parmesan or Romano.)

Mexican-Style Soup

Per pint of Basic Vegetable Soup, add:

½ cup water or stock
½ cup corn, fresh or frozen
½ teaspoon ground cumin
1 teaspoon fresh oregano leaves or ½ teaspoon dried oregano
1 small hot chili pepper, fresh, dried, or pickled; minced

Simmer for 5 minutes. Stir in:

1–2 teaspoons vinegar or the juice of half a lime

Top each serving with:

1 tablespoon yogurt or sour cream
1 small handful of fresh cilantro leaves, torn

Vegetable Stew

Per pint of Basic Vegetable Soup, add:

2 cups lightly packed chopped leafy greens
1 cup other diced vegetables (broccoli, kohlrabi, sweet peppers, etc.)
½ cup chopped mushrooms
¼ cup dried tomatoes, rehydrated for 15 minutes in ½ cup boiling water
   and minced, or 1 tablespoon tomato paste plus ½ cup water or stock
1–2 teaspoons fresh or ½–1 teaspoon dried herbs (Thyme is especially
   good. Use half as much rosemary or sage, which are very strong.)

Method #1

1. If using leftover Basic Vegetable Soup, reheat the leftover soup over low
   heat.

2. Meanwhile, cook the greens, vegetables, and mushrooms separately till
   tender then add to the soup. (If you have some already cooked leafy

greens on hand, as I recommended in the last chapter, you can simply add them when you add the rehydrated dried tomatoes.)

(*Tip:* You can put the vegetables in water in a pot, the greens and mushrooms in a steamer colander over the pot. Cover and steam until tender. If the greens and mushrooms are tender before the vegetables boiling below them, remove the steamer basket or colander and cover the pot. Yes, the flavors will cross over between the different ingredients, but since they are all going into the same stew anyway, that's okay.)

3. Stir in the tomatoes with their soaking water and herbs, cover, and simmer for 5 minutes for the flavors to "marry." Taste and add salt if needed. (The reason I don't just dump in the raw veggies and cook them in the already prepared soup is that the original vegetables in the soup are already perfectly cooked. They would be mushy by the time the newly added vegetables were done.)

Method #2
If making the Basic Vegetable Soup from scratch but heading straight for the stew variation, add the mushrooms to the carrots, onions, and celery. Add the dried tomatoes when you add the garlic. Add the raw diced vegetables and the tomato-soaking water when you add the potatoes or turnips. Add the raw leafy greens and the herbs for the last 10 minutes the soup cooks. Taste and add salt if needed.

Omnivore Soup

Any leftover cooked meat can be chopped and added to the Basic Vegetable Soup. Leftover cooked chicken is especially good in the Mexican-Style Soup above. You can also cook some local sausage in a frying pan until browned and then add that to the soup.

Cleanup

To paraphrase Chef Alton Brown, for every act of cookery there is an equal or greater act of cleaning up. Make peace with that fact. If you've got family and friends eating with you, get them to help with the dishes. If you're solo, clean as you go so that it's not as overwhelming when you're done with your meal.

Fill dirty dishes with water if they're going to sit in the sink for more than a few minutes—it makes the washing later much easier. Play some music during the cleanup, preferably something energetic. Don't bother drying dishes unless you need to put them away immediately; just leave them in the dish rack or dishwasher (with the door propped open) to air dry. When everything is clean, the cooking adventure is done for the day. Congratulate yourself on a job well done.

IF YOU DO JUST ONE THING...

One Soup and Seven Variations

Try this simple but satisfying soup and one or more of its variations. Soup is the great "use up" recipe of a locavore's kitchen because it can incorporate any seasonal vegetable. Also, it is one of the few foods that is actually better the day after you make it, so leftover soup is a good thing!

The measurements for the variations are per pint of the basic soup, making it easy to eat up your leftovers without getting bored. But you could also start each variation by making the whole recipe for the basic soup from scratch.

9

Simple Food Preservation

Fun, easy, safe, and deliciously addictive—those are the words that pop into my mind when I think "food preservation." But for many people "food preservation" conjures up exhausting images of some forebear sweating over vats of tomato sauce or scary thoughts of botulism.

To a twenty-first-century locavore, food preservation is not about putting up enough food to survive the winter. There is food available at the farmers' markets year-round, and some CSAs offer winter shares. It's about variety and pleasure and color and keeping it interesting until the next fresh crops start to come in. It's about nutrition, too, because a varied diet delivers more vitamins and minerals than a monotonous one. It's about pure convenience and benign laziness because food preservation equals a pantry of ready-to-eat foods. And for all those reasons, it's about being able to stick to a commitment to local farmers and local foods even during the "wolf months."

For example, here is my life in winter with food preservation: I get started making a pasta dinner. I add onions, cheese, garlic, and kale from the farmers' market, as well as dried tomatoes, basil oil, and maitake mushrooms from my pantry. My apartment is starting to smell seriously good. Here is my life in winter without food preservation: Get going on a pasta dinner, adding onions, cheese, garlic, and kale from the farmers' market. Okay, but just okay.

I first got into food preservation as a hobby. It was fun to be able to give my friends and family charmingly labeled jars of my jellies and pickles and enjoy their awed response—"You made this?" But as I got serious about local food, I discovered that food preservation not only added variety to my winter diet but also kept costs down and made my life easier when I didn't feel like cooking.

Preserved foods can create combinations that would be out of the question with fresh ingredients. For example, I can combine fresh pears (a fall crop) with blueberries that I froze when they were in season during the summer.

Do preserved foods contradict the idea of eating seasonally? No. The tomatoes, zucchini, eggplant, and peppers in my ratatouille were picked at the peak of their season, as were those blueberries in the freezer. I'm not settling for some inferior industrial agriculture crops bred for shipping sturdiness and shelf life. And I'm not adding to my carbon footprint by buying those ingredients out of season when they have to be shipped from other climates. What I am doing is adding variety to my winter fare.

Every culture has used food preservation to make their meals more interesting and to increase the nutritional value and health benefits of their meals. Vestiges of this still exist in that dill pickle that comes with a deli sandwich, or the chutney served with a curry. Some preserved foods, particularly lacto-fermented ones (see below) actually provide health benefits that their fresh counterparts lack.

What Makes Food Preservation Safe?

The bacteria that are dangerous to eat are also finicky. In order to survive, they need moisture, a very limited temperature range, and an even more limited pH range. So if we create an environment that is too dry, too hot, too cold, too acidic, or too salty, bacteria can't survive. And that's good news for us, because it makes it easy to safely preserve food.

To Sterilize or Not to Sterilize

Many food preservation recipes begin by telling you to sterilize jars. That's exactly the point at which, when I teach food preservation, I can see students' eyes glazing over. "Sounds complicated. If I don't do it right, I could die, right?"

Actually it is not that big a deal, and you don't have to sterilize everything for every food preservation recipe. Here are the rules: All lacto-fermented foods and those preserved in undiluted drinking alcohol or vinegar should be in clean jars, which means jars washed with dish soap and hot water. For foods to be processed in a boiling water bath or pressure-canned (see below), any jar that will be processed for 10 minutes or more need only be clean, not sterilized.

You may be thinking, "Well, heck, I'll just process everything for 10 minutes or more and never have to worry about the sterilization thing." But there are a few instances in which the extra time it takes to sterilize the jars (made up for by the shorter, usually 5-minute, processing time) is worth it. Pickles can lose their crunch if processed for too long, and jellies can lose their gel. Also, every minute of processing time equals a few more heat-sensitive nutrients lost.

To sterilize glass jars, place them in a large pot and completely cover them with water. Bring to a boil over high heat and start timing when the water reaches a full rolling boil. Boil for 15 minutes. If the jars are done sterilizing before you are ready to fill them, you can leave them in the hot water (once the heat is turned off) for up to 1 hour. Do not boil canning lids, as it can cause them to lose the adhesiveness of their seal. However, you can throw them into the water to scald them after the heat has been turned off.

Freezing

One of the easiest ways to preserve food is to use your freezer. Cooked leftovers like soups, sauces, and casseroles can be put in freezer containers (ziplock bags, plastic containers, or mason jars) and frozen for fast meals later on. Many fruits and vegetables can be frozen without any special preparation. Some, however, need to be blanched (dipped in boiling water) before freezing. This is because while your freezer is cold enough to prevent harmful bacteria from living, it is not cold enough to halt enzymatic processes. Those enzymes continue the decaying process of some raw foods even at freezing temperatures. Translation: If you put raw spinach into your freezer, you will have black goo when you thaw it out.

Foods that need to be blanched before freezing include all green vegetables, cauliflower, corn (on the cob or off), summer squash, and snap beans. To blanch vegetables, get a big pot of water boiling. Stir in your vegetables and start timing. For leafy greens such as chard, 1 minute in the boiling water is all you need. Heftier vegetables such as green beans and broccoli should be left in the boiling water for a full 2 minutes. Once the time is up, drain the vegetables in a colander and immediately run them under cold water or plunge into a bowl of ice water. This cold bath stops the cooking process. Once cool, drain well. If you are blanching greens, squeeze them really hard to get out as much liquid as possible. Pack into freezer bags or containers, and freeze.

Drying

Dried apples, apricots, and plums are among my favorite grab-it-and-go snacks. I've seen dried fruit for sale at farmers' markets on the West Coast, but the farmers here in the Northeast aren't offering any yet. So I make my own.

I also dry some vegetables, mushrooms, and herbs. One of the advantages of dried foods is that they take up very little space and can be stored at room temperature indefinitely.

Drying is also a great way to avoid throwing out food. When I can't make it through all the apples that came in my CSA share, or that bunch of herbs that is way more than I can use up before they spoil, I dry them to have on hand later in the year.

Do you need a dehydrator or other special equipment? No, but I do recommend a dehydrator if you want to dry foods in quantity—it makes the whole process much easier and more foolproof than air- or oven drying. It also uses less fuel and doesn't overheat your home when you want to dehydrate food during the hot months of peak harvest season.

In dry, hot Mediterranean climates, sun drying is an option (think sun-dried tomatoes). Unfortunately, while our summers are hot enough, they are also way too humid for drying food outside in the sun. The three methods that we can use to dehydrate food are air-drying (for herbs), oven drying, and using a dehydrator.

To dry herbs, secure the stem ends of 8 to 10 sprigs of any leafy herb with a rubber band. Do not use string—the stems shrink as they dry, slip out of the string, and fall on your floor. Hang the herbs away from direct light or heat. I have thumbtacks about 6 inches apart all along the edge of a bookshelf in a poorly lit part of my apartment, and they work perfectly well as herb-drying hooks. In a week or two the herbs should be crispy-dry.

If your home is very humid, you may need to finish drying the herbs in the oven. To do this, place the bundles of herbs on a cookie sheet in a low (125°F–150°F) oven for 5 minutes. If your oven doesn't have a setting that low (many don't), prop the oven door open with a dishtowel or the handle of a wooden spoon. You want the food to dry, not cook. Take the herbs out of the oven. They will crisp up as they cool, just like cookies do.

Crumble the leaves off the stems into glass jars (roll a sheet of paper into a funnel shape and place in the jar to make it easier to get the herbs into your jars and not all over your counter). Compost the stems.

The seeds and flowers of herbs can be dried in paper bags.

To dry fruits, mushrooms, and vegetables, use the oven or a dehydrator. Chop or slice them no thicker than ½ inch for drying. There are two extra steps you can take if you want a superior finished product: (1) Blanch vegetables in boiling water for 2 minutes and drain before drying. (2) Drop your fruit as you chop it into acidified water (a splash of cider vinegar in a big bowl of water will do the trick). Neither of these steps is absolutely necessary, but the produce will retain its color better if you spend the extra time.

For both the oven method and the dehydrator, your finished product should have a texture somewhere between leathery (for dried fruit) and crunchy-crisp (for vegetables and mushrooms). Err on the side of too dry if you're not sure whether the food is dehydrated enough. If there is too much moisture left in the food, it can mold in storage. Properly dried, dehydrated food can be stored in glass jars for years.

The oven method: Arrange your chopped produce in a single layer, making sure none of the pieces are touching, on a cookie sheet or, better, a rack in a 125°F–150°F oven (see note about low oven temps above). Drying time will vary depending on what you are drying. For example, mushrooms only take a couple of hours, but carrots can take as long as overnight. If using a cookie sheet, turn the food a few times during the first couple of hours to avoid sticking.

With a dehydrator: Preparation for drying food in a dehydrator is the same as for oven drying: Slice no thicker than ½ inch, with the optional steps of blanching for vegetables or putting fruit into acidified water. Most dehydrators come with stackable trays that have holes in them to allow the air to circulate around the food. Mine is a round floor model only 1 foot high and about 1½ feet in diameter. It has temperature settings that range from 95°F to 155°F. The temperature I use for most fruits and vegetables is 135°F.

To dry greens, preheat oven to 250°F. Wash and dry the leaves, removing any thick midribs. Tear into approximately 2-inch pieces. Spread in a single layer on a cookie sheet. Bake for 15 to 30 minutes depending on how thick the leaves are (kale may need the full 30, spinach only 15). The leaves will be deep green and crispy when they come out, but not burnt. You can use the crispy green "chips" as a garnish or store them to reconstitute later for soups and dips. Kale is by far the best green for drying.

To reconstitute dried veggies and mushrooms, pour boiling water over them and soak for 15 to 20 minutes. Save the soaking water to add to soups and sauces.

What do I dry most? Apples, plums, pears, and apricots because they make great snacks; tomatoes because I love the taste of sun-dried tomatoes and the oven-

or dehydrator-dried ones taste just as good; celery and carrots because they are useful in soup stock and other recipes; mushrooms because they reconstitute beautifully.

Pickling

Almost any vegetable can be pickled, and so can fruits. (If you think pickled fruit sounds weird, you have got to try pickled cherries with some of Old Chatham's Hudson Valley Camembert. My friend Anne discovered this combination, and it is delectable. Guess that's why this is the second time I'm mentioning it!) There are two main kinds of pickling: pickles preserved with vinegar and those preserved by lacto-fermentation. Both methods preserve food in brine that is either too acidic or too salty for harmful bacteria to survive.

In all pickling recipes, and indeed any preserved food recipe, be sure to use sea, kosher, or canning salt. Do not use iodized table salt—it can discolor your final product.

Unless you are blessed with your own unchlorinated water source, use filtered water, especially for lacto-fermented foods. The chlorine in municipal tap water can prevent successful fermentation.

Preserving in Vinegar

Vinegar is the most common pickling brine. Simple pickle recipes can be as easy as covering the food you are preserving in vinegar (see the recipe for preserving hot peppers below). More complicated recipes for things like bread-and-butter pickles add spices, water, and sugar or honey to the vinegar. There are only two rules you need to remember in order to safely use vinegar to preserve foods:

1. Always use vinegar that is at least 4.5 percent acetic acid (this includes most commercial vinegars, and the acetic acid strength is always listed on the label). If you want to use homemade vinegar, you need to test it to be sure it is acidic enough for safe preserving. See the Useful Resources appendix for where to get a testing kit and instructions.

2. If diluting the brine to make a milder tasting pickle, make sure you have at least 50 percent vinegar to water. Less than a 50–50 ratio isn't acidic enough to preserve food.

Lemon Herb Vinegar

This is my answer to the fact that there is no citrus grown in the Northeast because the trees can't survive our winters. I confess that since my 250-mile diet year ended, I do allow myself the occasional lemon or lime. But most of the time I use lemon herb vinegar. This vinegar is imbued with the flavors of lemon-scented herbs. The herbs provide the citrusy flavor, and the vinegar provides the sourness. I won't pretend the effect is identical to a fresh lemon, but it is equally good in its own way.

The recipe is simple: Loosely pack a jar with fresh or dried lemon-scented herbs (lemon thyme, lemon basil, lemon balm, lemon verbena, lemongrass, etc., all available at farmers' markets, or you can grow them yourself). Pour in enough vinegar (I recommend a cider or white wine vinegar) to cover the herbs. Cover the jar and leave the mixture to steep for two weeks. Strain out the herbs and pour the lemon herb vinegar into a glass bottle. Label, cap or cork, and enjoy. For a stronger lemon flavor, after you've strained out and discarded the herbs that were in the vinegar, add a fresh batch of lemon herbs to the jar and pour the already infused vinegar over them. Leave for another two weeks before straining and using.

137

Four Super-Simple Pickle Recipes

Cornichons

Cornichons are those tiny, tart cucumber pickles often served at restaurants alongside pâté, cheese, or cold cuts. They are incredibly easy to make, and you don't have to sterilize jars or process anything in a boiling water bath. You don't even need to use canning jars. You also don't have to make them all at once. You can add a few baby cukes at a time as they come out of the garden.

And that's the catch: You can't buy cucumbers small enough to make cornichons; you have to grow them yourself. They should be no bigger than your pinkie finger. This is a treat only gardeners and farmers can make, and they make a fabulous gift.

Small cucumbers, no bigger than your little finger
Non-iodized salt
Cider or white wine vinegar

1. Thinly slice off the blossom end of the cucumbers. The blossom end contains enzymes that turn pickles mushy. If you aren't sure which end is the blossom end, slice off both ends. Rub off any prickles.

2. Place the cucumbers in a non-aluminum container and cover with salt. Leave in the refrigerator for 24 hours.

3. Remove the cucumbers and brush off excess salt (it's okay if a little salt clings to them). Place in a clean glass jar and cover with vinegar. Cover and return to the refrigerator. They are ready to eat after two weeks and keep in the refrigerator indefinitely, although they start to lose crunch and quality after a year. If you like, you can add garlic cloves, peeled baby onions, peppercorns, or bay leaves to the vinegar.

Hot Peppers in Vinegar or Sherry

Use any kind of chili pepper for this. Slice larger peppers into rings, removing the seeds. Prick smaller peppers with the point of a paring knife to allow the vinegar to penetrate. Place in a glass jar and cover with vinegar. In two weeks they are ready to use and will keep, refrigerated, indefinitely. Actually they will even keep at room temperature, but I find they start to lose their bright colors when stored that way. Use the pickled peppers in any recipe that calls for jalapeños or other hot peppers. When you've used up all the peppers, the vinegar makes an excellent hot sauce similar to Tabasco.

My friend Rob makes a variation of this recipe using sherry instead of vinegar. The peppers can be used as above. The sherry that's left over after you've eaten the peppers is a wickedly good addition to sauces.

24-Hour Refrigerator Pickles
MAKES APPROXIMATELY 2 PINTS

1 cup water

1 cup cider or white wine vinegar

3 tablespoons kosher or non-iodized salt

2 tablespoons honey

2 large cucumbers, cut into spears

2 small cloves garlic, peeled

2 sprig of dill leaves or 1 dill flower head

½ teaspoon red pepper flakes

1. Bring water, vinegar, salt, and honey to a boil. Meanwhile pack two clean glass pint jars with the cucumbers, adding one clove of garlic, two sprigs of dill, and a pinch of red pepper flakes to each jar.

2. Pour the hot brine over the other ingredients. Secure lids and store in the refrigerator for at least 24 hours before eating.

Chutneys

Chutney is a sweet-and-sour pickle of finely chopped fruits and/or vegetables. A catsup, or ketchup, is just pureed chutney, so the red stuff on your burger is technically pureed tomato chutney. You can use almost the same brine as in the refrigerator pickle recipe above, leaving out the water and increasing the amount of honey ½ cup.

Add a mixture of finely chopped vegetables and fruit, including some dried fruits if you like. Green tomatoes and apples are an especially good combination. Bring to a boil; reduce heat and simmer. Continue to simmer, stirring frequently, until the ingredients are tender and the chutney is as thick as you want. Store in the refrigerator or process in a boiling water bath for 10 minutes. Experiment with other seasonings including mustard seeds and ginger (wild ginger, or Asarum, if you're keeping it all local).

Serve chutney alongside curries, mixed into rice dishes, or with any meat.

Lacto-Fermentation

Lacto-fermentation is a pickling process that relies on natural fermentation to preserve the food rather than the addition of vinegar. Lacto-fermentation is the way real kosher dills and sauerkraut are made, and the process creates a much lighter, fresher taste than vinegar pickles.

In the first stage of lacto-fermentation, salt creates an environment that is too alkaline for harmful bacteria to survive. Fortunately there are beneficial bacteria (*Lactobacilli*, the same genus of healthy bacteria that turn milk into yogurt) that can survive in a salty environment. The beneficial bacteria take over and start fermenting the vegetables. They are also incredibly good for you. Lacto-fermented foods have higher nutritional content and are easier to digest than their nonfermented counterparts. They promote good digestion and even contain anticarcinogenic substances.

Lacto-Fermented Snap Beans
MAKES 2 PINTS

1 pint filtered water (Do not use chlorinated tap water—it interferes
   with the fermentation process.)

1 tablespoon non-iodized salt

1½ pounds snap beans, washed and stem ends removed

1. Dissolve the salt in the water.

2. Pack the beans tightly into clean glass jars. Trim them as needed to allow
for just over 1 inch of space between the tops of the beans and the rim of
the jar.

3. Pour the brine over the beans, leaving 1 inch of headspace. Cover and
leave at room temperature for three days. Transfer to the refrigerator and
wait at least a month before using.

This recipe can be used with almost any vegetable. Swiss chard ribs or thinly sliced raw turnips are especially good.

During the first few days of fermenting at room temperature, the liquid will get bubbly and smell a little musty. As the vegetables continue to ferment in the refrigerator, they develop a clean, lightly sour smell. Discard any that look cloudy or smell "off."

You can eat your lacto-fermented vegetables like pickles straight out of the jar. To serve as a main dish rather than a condiment or garnish, soak the vegetables in cold water for a few hours first to leach out some of the salt. Lacto-fermented vegetables are delicious chopped fine and added to grain- or bean-based dishes and soups.

Preserving in Salt

While salt provides the first step of lacto-fermentation, it can also be used to preserve food without fermentation. The recipe below is great for preserving odds and ends from the garden or from your CSA share. You can vary the vegetables you use, as long as you always keep the ratio of no more than four parts minced veggies and herbs to one part salt.

Verdurette

1 part leafy greens

1 part root vegetables, especially carrots (not potatoes, which don't add much flavor, or beets, unless you want your verdurette to be magenta colored)

1 part alliums (onions, garlic, leeks, shallots, etc.)

1 part fresh herbs

1 part non-iodized salt

1. Finely mince a combination of the above (a food processor makes quick work of this).

2. Measure the minced herbs and vegetables and stir in four parts non-iodized salt (e.g., 4 cups minced veggies to 1 cup salt).

3. Pack the verdurette into a glass jar and cover. It will keep in the refrigerator indefinitely. Use your verdurette to flavor dips and soups, leaving out any salt called for in the recipe.

Preserving in Oil

Most vegetables can be preserved using the recipe below, which combines vinegar pickling with preservation in oil. The vinegar creates bacteria-killing acidity and adds a slight tanginess to the taste. The oil keeps air from reaching the vegetables and also picks up the flavors of the fresh herbs.

Marinated Vegetables or Mushrooms

This recipe works especially well with eggplant and fresh mushrooms, but you can use it with other foods as well. If using a hard vegetable such as cauliflower, steam or boil the chopped vegetable first for 2 minutes before proceeding as below.

1. Bring vinegar to a boil in a non-aluminum pot. Add the vegetables or mushrooms. Be sure there is enough vinegar to cover them. Boil in the vinegar for 5 minutes. Drain the vegetables in a colander or sieve. (You can reuse the vinegar for another purpose if you like.)

2. Loosely fill a glass jar(s) with the vegetables. Tuck in sprigs of fresh or dried herbs (thyme, marjoram, rosemary, and bay work especially well). Cover with extra-virgin olive oil (not local, I know; this recipe is not for purists). Gently press down on the vegetables with a spoon to release any air bubbles. Make sure the vegetables are completely immersed in the oil. Cover and store in the refrigerator for at least two weeks before using so that the flavors have a chance to develop.

3. Serve at room temperature with good, crusty bread.

Basil or Other Herb Oils

Some herbs, including basil, sorrel, cilantro, and parsley, do not dry well. They don't freeze well either. But you can preserve the bright flavors and vibrant green color of these herbs by creating herbal oils and then freezing them.

Basil Oil Cubes

MAKES APPROXIMATELY 14 CUBES/1 ICE CUBE TRAY

1 large bunch of basil
1 pot of boiling water
1 bowl of ice water
1½ cups extra-virgin olive oil
1 ice cube tray

1. Holding the bunch of basil by the stem ends, swirl the leaves in the boiling water for a few seconds. Immediately plunge into the ice water. This blanching preserves the herb's bright green color when frozen.

2. Remove the blanched leaves from the stems and put them into a blender along with the olive oil. Puree until well blended with a few flecks of leaves still visible.

3. Pour into ice cube trays and freeze. Pop the frozen basil cubes out of the trays and store in freezer bags or containers.

Each cube is approximately 1 tablespoon of basil oil, so you can take out just what you need. Use basil oil to top soups, as a basting sauce for chicken, as a dipping sauce for bread, or to make pesto.

144

Canning

If you open the cabinet above my kitchen sink, you'll see a colorful abundance of jarred food. Light orange peaches stand beside deep red tomatoes and green and yellow pickled snap beans. The advantage of home-canned foods is that because they are vacuum-sealed, they can be stored at room temperature, which opens up freezer and refrigerator space. There is also tremendous satisfaction in seeing those rainbow jars filled with food that was preserved at the peak of the harvest.

If you are one of those people who thought "botulism" when you heard "food preservation," here is what you need to know. There are two distinct canning methods: the boiling water bath and pressure canning. The boiling water bath requires no special equipment but is only safe for specific foods. Pressure canning requires special equipment and a bit of a learning curve, but it can be used for any food. The only way you can get botulism from home-canned food is if you use a boiling water bath for foods that actually need to be pressure-canned. Once you know which foods can safely be processed in a boiling water bath and which can't, you are guaranteed safe.

The only foods that can be safely processed in a boiling water bath are:

- All sweet fruits
- Vinegar pickles
- Jellies and jams made with sugar or honey
- Tomatoes (with added vinegar, citric acid, or lemon juice to up their acidity)

Lacto-fermented foods can also be safely processed in a boiling water bath, but doing so kills all the health-promoting bacteria that are one of the major reasons to eat lacto-fermented foods. Don't do it.

Boiling Water Bath Canning for Sweet Fruits, Tomatoes, Pickles, Jellies, Jams, and Chutneys

First off, get yourself some canning jars with two-piece lids. With these and a boiling water bath, you can create a vacuum-tight, air-proof seal.

Please read the above information about what can and cannot be safely processed in a boiling water bath. The actual process is very simple. You'll need:

A pot tall enough for your jars to be covered with 1–2 inches of water above their tops
A rack that fits into the pot or a dishtowel

That's it. Place the rack or dishtowel in the bottom of the pot. Place your jars of fruit or preserves in the pot. Fill with enough water to cover your jars by at least 1 inch. Turn heat to high. When the water comes to a full boil, start timing according to the recipe you are following, e.g., "process in a boiling water bath for 10 minutes." Check the recipe, a reliable canning book, or an online guide (see Useful Resources) for how long each kind of food needs to be processed.

If you are using a dishtowel instead of a rack, you need enough jars in the boiling water bath to completely fill the bottom of the pot. (If you only have one or two jars, the towel will float and the jars will tip over. You want the jars securely upright during the processing.) To can just a couple of jars, fill a bunch of empty jars with water and place those around your jars of food.

When processing tomatoes in a boiling water bath, add a teaspoon of vinegar to each jar. Some tomato varieties are acidic enough to safely process without this, some may not be. Adding vinegar is your insurance for the varieties that are not.

When the time is up, remove the jars and let cool, undisturbed, for at least 8 hours. The undisturbed part is because jostling could undo the seal on a sealed but still warm jar. My practical experience says that you can move them if you are very, very careful.

As the jars cool, the lids will seal. You'll hear a ping or click as each jar seals—one of the more satisfying experiences of canning, though darned if I can say why. When I have a batch of jars cooling, even if I am busy doing something else, I mentally count off the sealed jars: "That's one!" "That's two!" I used to think I was the only one who did this until I visited a friend and fellow canning enthusiast. We canned about a dozen jars of blueberries, and while they were cooling and the lids were pinging, her husband (who is not into food preservation) called out, "That's one! That's two!"

The lids of jars that are successfully sealed are slightly concave and solid versus the slightly convex and flexible lids of unsealed jars. This sounds subtle, but it's really not. Once you've felt the difference between a sealed and an unsealed canning lid, you'll know it forever. If this is your first time canning, I recommend pushing on the lids of the jars before you put them into the boiling water bath so that you'll know the difference in how they feel once they seal. However, do not push down on the lids while they are cooling—this can create a false seal.

Note: You can reuse the screw band rings that are part of two-piece canning lids, but you are not supposed to reuse the lids themselves. The reason is that the adhesive ring on the lids can lose its sealing ability with reuse. I confess that I do sometimes reuse the lids, but I am taking a gamble because there is a chance that the jars won't seal. Until you are very, very sure you know how to tell a sealed jar from an unsealed one, use new lids every time.

Pressure Canning

Pressure canning requires special equipment but has the plus of enabling you to process pretty much any food in existence. The reason is that the pressure creates temperatures that are even hotter than boiling water, blasting potentially dangerous bacteria out of existence.

I'm not going to go into detailed pressure-canning directions here, because every pressure canner is slightly different and comes with its own instructions. My own pressure canner is the All-American brand, which I highly recommend.

I don't pressure-can as many foods as I used to. I find that sometimes the quality of the finished product is not great, and I know that many vitamins are destroyed during the process. What do I mean when I say the quality is not great? I mean they're the homemade equivalent of those mushy, bland canned vegetables you may remember from childhood. (Did I just give my age away?)

There are, however, three reasons my pressure canner remains in frequent use. One is for pressure-canning stock so that I can store it at room temperature when my freezer is too packed to have room for it. The main nutritional value of the stock is its minerals, and minerals are not destroyed by the pressure-canning process the way vitamins are. I also can ratatouille—a summer-in-a-jar mixture of eggplant, squash, sweet peppers, and other vegetables that must be pressure canned. Yes I lose some of the vitamins that way, but ratatouille is one of my favorite winter convenience foods, so it's worth it.

The other way I use my pressure canner is as a boiling water bath. Using the canner requires less water than using any other large pot, and that cuts out a lot of the waiting-for-the-water-to-boil time. Here's how you do it:

Using Your Pressure Canner as a Boiling Water Bath

To use a pressure canner as a boiling water bath, place the rack that came with the canner in the bottom. Add jars of food that can be safely processed in a boiling water bath (see above), and add water to come up to the shoulders of the jars. Put the canner lid on and fasten the screws, but leave the vent open. Turn heat to high. Once steam is coming out of the vent at full blast, start timing according to the recipe's instructions. Give the canner 2–3 minutes to cool down before taking off the lid and removing the jars.

Optional but Useful Canning Equipment

A canning funnel makes filling jars much easier. It also has household uses beyond canning, such as filling jars with dry beans or grains.

A jar lifter is very handy for securely getting jars out of the hot water after processing them and safely transporting them to wherever you are going to let them cool.

Where Do I Get Canning Supplies?

Most hardware and household goods stores carry canning jars and lids. You may have to do a bit of translation. The first time I asked for canning jars at my neighborhood hardware store, I got a blank stare in response. I started describing what I wanted and the guy said, "Oh! You mean jelly jars. Sure, I've got those."

Pressure canners, jars, lids, jar lifters, canning funnels, and other equipment can be ordered online (see Useful Resources).

Now What Do I Do with It? Making Use of Your Locavore's Pantry

When I first got into food preservation, I merrily made things just because they sounded cool. Green tomato chutney, rhubarb-strawberry compote, spicy carrot pickles, Juneberry jam. I had all these jars of food, and really, just how many pickles can one gal eat? How many pieces of toast with homemade jam? At first it seemed that some of my homemade preserves might go to waste because they were more in the condiment than mainstay category.

Since then I've found ways to use everything I've made and wish I'd made more. Some of the ideas for how to use them came from browsing through old-time books and recipes from back in the day when home-preserved fruits and vegetables were a familiar part of everyday meals. Some I dreamed up on my own.

I assume you already know about jam on toast and a pickle to go with your sandwich. Here are a few other ideas:

- **Jam or other fruit preserves with yogurt.** This is basically what commercial fruit-flavored yogurts are, but they don't tell you that.
- **Fast dips and spreads made with chutney**. Take any kind of homemade chutney and add it to yogurt for an instant dip. Your call whether to blend it smooth or leave it chunky.
- **Instant ketchup.** The ketchup we're familiar with is actually just pureed tomato chutney. You can make a similar-tasting (but better) condiment out of any home-made chutney by pureeing it.
- **Chopped vegetable pickles added to bean soup.** In my family, it was standard to add a splash of vinegar to lentil or other bean soups. Adding chopped pickled carrots or other vegetables contributes the same light spark of sourness to balance the blandness of the beans and adds interesting texture and color as well.
- **Pickled hot peppers** can be used in place of fresh or dried hot peppers (jalapeños, etc.) in any recipe.

For more ideas on how to use your preserved foods, see the useful substitutions section on page 112.

see the useful substitutions section on page 112.

If You Do Just One Thing . . .

Foods that can be frozen without blanching or any other special treatment

- *Berries and other fruit*
- *Sweet and hot peppers*
- *Onions*
- *Tomatoes (not for salads—they'll never be fresh again—but grand for sauces and soups)*
- *Meat, poultry, fish; raw or cooked*

Tip #1: Spread berries out on a dish or cookie sheet and freeze. Then pack the frozen fruit into freezer bags or containers. The individual berries will remain loose rather than turning into a big frozen clump, and you'll be able to scoop out just what you need.

Tip #2: Foods that will be chopped in most recipes, such as onions, should be chopped before freezing. Use the same cookie sheet method described for berries, above, so that you can take out just what you want to use each time.

10
Feasting for Free

Wild edible plants and mushrooms are the quintessential local foods. They're growing right under your feet, and they're free.

"Why do you forage?" a student asked me at a wild edible plants tour I recently led in Brooklyn's Prospect Park. "Is it the adventure or that the food is so much better?"

I started foraging as a toddler in San Francisco's Golden Gate Park. My great-grandmother was still alive. She had grown up in a village on a small Greek island where going out to pick the first leafy greens of spring wasn't "foraging," it was a chance to score a gourmet treat. So one answer is, "I've always done it."

But why do I keep doing it? Yes, the adventure is part of it. Foraging is a treasure hunt. Today, for instance, while leading a foraging class we found a huge patch of black raspberries that I'd never stumbled across before. They weren't quite ripe. I marked the location down on my iPhone map and am definitely going back in a couple of weeks, when the berries will be perfect. Adventure as well as good eating was part of the fun when my friends Mark and Ellen stumbled onto a watercress bonanza. Mark had said they were going to find watercress that day, and they did find a small patch of it. Then they walked over a slope and around a corner, and there before them spread more watercress than they could ever eat. They were kind enough to share their discovery with me, and we've revisited that watercress patch many times since.

The adventure of wild foods heightens my awareness of place and season, especially when I factor in wild edible mushrooms. A certain mushroom has to look exactly like this and only grow on these kinds of trees and only in autumn, for example. I have to pay attention instead of tuning out my surroundings the way I too often do when I walk down a crowded street.

Another reason I forage is that it's free food. Knowing how to identify and collect wild edible plants lowers my locavore food bill.

Wild foods are often more nutritious than standard crops, even when compared to organically grown fruits and vegetables. Many wild edible plants contain higher levels of vitamins, minerals, and phytonutrients than domesticated plants. Because of that, they also often have stronger flavors than domesticated plants. Over the centuries, humans bred many food plants to increase blandness, sweetness, and bulk. The trade-off is fewer nutrients. High-yield crops are typically lacking in the micronutrients their un-tampered-with counterparts contain. Those same nutrients are what give wild edibles their pungency.

Another reason I love foraging is the chance to cook with ingredients that I can't get any other way. When was the last time you saw Juneberries for sale? Or garlic mustard seeds, daylily buds, tender chickweed salad leaves, or butternuts?

There is also a baseline sense of security that I get from being a forager. If I'm on a train or in a car passing through terrain I've never visited, I have a habit of mentally checking off the plants I see outside the window as we zip past. *There's burdock,* I might think, *and evening primrose, wild carrots, and elderberries. I'd be okay here if the train stopped and wasn't moving again anytime soon.*

Not that long ago, knowledge of a few wild edible plants was common, whether or not they were something a person ate every day or even wanted to eat every day. During World War II, for example, when supplies ran short in England, many people foraged in the countryside to supplement their rations. They may not have been happy about it, but they knew how to, which is something that can't be said about most people I know. There is something reassuring about knowing that there is free food growing all around you, even if you decide not to collect it that day.

Foraging Fills in the Gaps

You'd almost think someone had scheduled it on purpose: At the very end of winter, just when the farmers' markets lack any fresh produce and are running out of the crops stored from the previous year, the wild edible plants season kicks in. And just when the wild edibles season hits a lull in midsummer, the agricultural season is in full swing. Wild edibles and domesticated crops fill in the gaps in each other's calendars.

For example, in March any garlic you have left from last year, and any at the market, is probably an already sprouting, slightly bitter affair. But March is peak season in our area for wild field garlic. By late spring the farms are offering garlic scapes and green garlic, but the wild stuff is dying back. The main garlic crop gets harvested in July and keeps for a few months. The wild garlic crop kicks back in late fall and lasts through winter into the start of the following spring.

Survival versus Gourmet Foraging

Some people show up at my foraging classes expecting a how-I'd-survive-if-I-were-lost-in-the-woods tutorial. Yeah, I can teach that, but it's not the kind of foraging I do myself. That kind of survival foraging is really a worst-case scenario. Being stranded in an unfamiliar landscape and setting off hoping to find something edible to fend off starvation is not how indigenous peoples and hunter-gatherer societies forage. In those cultures there is advanced knowledge of what is available when and where, and the foraging is very efficient.

You wouldn't go looking for fiddleheads in August in a sunny spot, for example, because fiddleheads are an early-spring wild vegetable that prefers to grow in the shade. You would beeline for the white oak grove in autumn because white oaks have particularly large and not too tannic acorns that are ripe in autumn. Part of the fun of foraging is getting to know the immediate landscape and the seasons intimately—to remember where I found that white mulberry tree last June and then in the fall head straight for that particular oak tree where I collected hen of the woods mushrooms two years in a row.

Do I also have an overly developed love of risk-taking (since some of those plants and mushrooms could be poisonous)? No!

The First Rule of Foraging: If in Doubt, Throw It Out

Always be 100 percent certain of your identification before trying a new plant or mushroom. Although some field guides and Web sites continue to suggest that chewing on something and then spitting it out and waiting to see if you have a negative reaction is a good way to test for edibility, it's a bad idea. And don't assume that because you see an animal or bird eating something that means it's safe for human consumption—birds and other wildlife have very different digestive systems from ours and often chow down on foods that we can't safely eat.

A good field guide is essential to cross-check your identification details until you are familiarized with a plant. Nowadays good field guide information can be found online as well as in books, although that won't help you when you're staring at a plant in the field and trying to decide if it is what you think it is (though I have been known to pull out my iPhone in the park and go online to double-check a plant ID).

You already have plant identification skills; you just may not realize that you do. Although I can't stress the importance of correctly identifying wild edibles enough, I also want to encourage you to get out there and forage. While it's essential to make sure you've got the right plant before you swallow it, that's much easier than many people seem to think.

When I lead wild edible plants classes, I often start off by pointing to a dandelion and asking, "What's that?" Inevitably the students know that it's a dandelion. "How do you know?" I ask. "The yellow flower," they reply. I point out that lots of plants have yellow flowers and ask how else they can tell that it's a dandelion. They point out the shape of the leaves. I point out other characteristics that they are observing but may not be consciously aware of, such as the way dandelions grow close to the ground in a rosette pattern and without the tall, leafed, branching flower stalks that other plants have.

You demonstrate your plant identification skills when you grab a bunch of parsley at the store without checking the sign, even though it is next to the cilantro, which looks similar. Or when you recognize chard versus kale versus lettuce. You already know how to identify plants—you just need to be introduced to a few new ones.

Not Just What, but When and What Part?

There are a few foraging identification skills you'll need beyond being able to match up the visual of a plant to its description in a field guide.

It's not always enough to know what a plant is. You also want to know when you can eat it and *what part* of the plant you can eat. Some plants like pokeweed are delicious in spring when the shoots first come up but become toxic as the plants mature. Others have edible fruit, but the rest of the plant is poisonous. And with still others, like the dandelion, all parts are safe to eat anytime but are so bitter at certain times of the year that you wouldn't want to.

Don't panic. These, too, are considerations that you (or your farmer) already take care of daily when it comes to domesticated crops. Tomato leaves are poisonous, for example, but obviously tomatoes are not. Lettuce becomes unpalatably bitter once it "bolts" (sends up a flower stalk). There's no mystery to learning what part to eat when; you've just got to learn it. You eat tomatoes, not tomato leaves, right?

One of the best tools you can use as a forager is your nose. Drawings and photographs are nice, but the fact that mint has a square stem and opposite leaves with toothed margins doesn't clue you in to what mint smells like. But once you've smelled mint, you know that smell anywhere. I am a big believer in scratch-and-sniff plant ID.

USE YOUR NOSE

There aren't many general rules for safe foraging beyond being sure of your identification, but here is one:

Anything that smells like onions or garlic is safe to eat.

There are poisonous plants that look similar to wild garlic and onions, but not one of them smells *like garlic or onions. If you think you've found a wild* Allium *(the plant genus that includes onions, garlic, shallots, and leeks), double-check your ID by breaking off some of the plant and giving it a sniff. If it has that garlicky smell, you're good to go.*

Know Your ID . . . But Don't Get Skittish!

Some of the people who attend my foraging tours are extremists. Either they get so excited about the idea of free wild food that they are ready to pop any plant into their mouths, or they are so terrified that they pull away from a taste I offer them (even though they signed up for a wild edible plants tour and an expert just identified it for them).

A case of the overly gung-ho happened on one foraging walk in New Jersey. I'd just finished explaining the identification particulars for a tasty edible called garlic mustard *(Alliara petiolata)*. I started walking toward the next delicious wild plant, a stand of Japanese knotweed *(Polygonum cuspidatum)* up ahead.

Behind me I heard one of the participants say, "This one looks a little different. Tastes different too." That is not something you want to hear if you are leading a wild edible plants class! I turned pale and ran back to see what he had just eaten. Fortunately it was a harmless violet leaf, which has a similar heart shape to garlic mustard. I learned my lesson—now I ask tour participants not to sample anything unless I've confirmed their ID.

That guy made his mistake because he forgot that when identifying a plant or mushroom, one characteristic match is not enough. Even two are not enough. The garlic mustard I'd shown him how to identify has heart-shaped leaves and grows in a rosette pattern, which means the leaves emerge from one central point on the ground. The violet leaves he mistakenly munched on are also heart shaped and also grow in a rosette arrangement. But violet leaves are a darker shade of green and have a much more pointed tip, sharper teeth along the margins (the edges of the leaves), and prominent veins on the underside. Also, they don't smell or taste like garlic. When identifying a plant for the first time, you want all the characteristics mentioned in your field guide to match.

An exception to this is height. A field guide might say that Japanese knotweed is 4 to 7 feet tall, but when it first comes up in spring, which is when it's tasty, it is only 1 foot or less high. Keep in mind that the heights given in books and online field guides often refer to the mature height of the plant. You may find yours when it is much smaller.

On the other hand, often there are people who follow the whole tour enthusiastically, take diligent notes, but then refuse to actually try anything. Two hours into a wild edible plants tour, I put some lemony wood sorrel leaves into my mouth and invite the tour participants to partake as well. Inevitably a few back away, their expressions a mix of fear and embarrassment because they realize this is one time they really shouldn't be afraid.

It's not just adults who are conditioned to mistrust food that doesn't come with a label. There are children who have grown up so used to packaged food that they don't even know how to recognize a blackberry.

I grew up in a city, but even so I learned how to recognize the blackberries in the parks and on the occasional countryside camping trip my parents took me on. A few years ago I was working for a few days in upstate New York for a dance company, Dances Patrelle. The kids in the show had traveled up with us from the city along with their parents and/or chaperones. The motel we stayed in was a little on the dilapidated side, and the only restaurant nearby was a pancake house.

One morning I spotted a glorious thicket of blackberry canes just outside the restaurant parking lot. The berries were perfectly ripe, just begging to be picked. *Wonderful!* I thought. *Everyone has been grumbling about the motel and the restaurant, and this will cheer them up.* I picked about a quart of blackberries and came back to the restaurant grinning and planning to distribute them among everyone's plates of pancakes.

Both kids and adults recoiled. "How do you know they're blackberries?" they asked. Only two people would share the fruit with me, along with stories about how they used to pick blackberries as a kid.

When did we give up our knowledge of how the land around us sustains us? When did we decide that a label stuck on a package by a stranger was more trustworthy than our own ability to recognize a blackberry?

Get out there and forage. Do it smartly, safely, and efficiently, but most important, do it.

Foraging Safely

- **It's worth saying again:** Be sure of your identification! When in doubt, leave it out. Get a good field guide or use the online field guides to confirm your identification until you are 100 percent positive. I've recommended my favorite field guides and Web sites in Useful Resources.
- **Be certain that the area you are harvesting** has not been treated with chemicals. Never collect leaves from a garden or farm that uses large amounts of chemical fertilizers. Many edible "weeds," especially those in the Chenopodiaceae family (including domesticated edibles such as spinach, beets, and chard), can store the nitrates from fertilizers in potentially toxic amounts.
- **Never harvest roots, leaves, or stems** near a heavily trafficked road. Flowers and fruits tend to collect fewer toxins from the soil, but they may still be affected by air pollution. I do collect these from street trees, but I wash them well before using. Plants collected uphill from a pollution source collect fewer toxins than plants downhill from it, so you can harvest closer than that 50-foot rule if you are uphill from the road. Alliums (wild garlic, etc.) and sunflowers are so good at absorbing

toxins from the soil that they are being used to clean the land near Chernobyl. Collect these even farther from sources of pollution than you do other genera.

- **The first time you eat any plant,** try just a small amount. Wild plants tend to have stronger flavors than domesticated plants and can cause stomach upset in people who are used to blander vegetables. Also, just as most people can eat shellfish but some are allergic to it, there is always the possibility of a food allergy. Once you have tried a small amount of a plant and know that it agrees with you, you can increase the amount that you eat in future recipes.
- **Know which parts of the plant are edible,** and eat them only in season. Many plants, such as tomatoes have one part that is edible while another is poisonous. Others, like poke, are choice edibles in season but toxic if eaten out of season.

Where to Find Wild Edibles

In some ways foraging is easier in the city than in the wild, because many plants love the disturbed soil created by human activity. In addition, there tends to be a greater diversity of trees and shrubs planted near one another than you would find growing naturally in a forested area. Following are the places where I do my urban foraging.

Parks

The first foraging I ever did was picking dandelion greens in San Francisco's Golden Gate Park with my great-grandmother. Because I've lived in cities for most of my life, much of my foraging has happened in parks. This is true for most urban foragers. There are some downsides, but first I want to talk about the upsides.

Parks are rarely sprayed with the herbicides and pesticides that botanical gardens and even home gardens often are. When park grounds have been treated with chemicals, clear signs are posted letting you know.

There are actually more varieties of wild edibles on offer in a city park than usually would show up on a walk through pristine woods. Many of these plants are "invasive aliens," which is to say not native to our area. They are prolific in parks, city lots, and urban gardens. You are not harming any plant population by harvesting invasive plants.

Most of us who teach wild edible plants in the city do so in the parks, but that brings me to an important question that always comes up:

Is It Legal?

I've spent hours online and on the phone trying to get the lowdown on whether foraging in the parks is actually illegal and, if it is illegal, what are the specifics of the law. I struck out. Finally I e-mailed "Wildman" Steve Brill, who is certainly the most well-known wild edible plants expert in the Tri-State area, regularly leads foraging tours in area parks, and who was once arrested for harvesting dandelions in Central Park. I asked him what he knew about the current laws pertaining to park foraging.

"It's a gray area. There's no law again foraging, but there is a Parks Department regulation against removing anything from the parks without permission. Even allowing toddlers to take home autumn leaves is verboten! In practice, if you're discreet no one will bother you, but you might get a warning from a park ranger if you're seen. In the fall Asian people collect ginkgo fruit for its seeds, and no one goes after them. Everyone knows they do this, so since it's allowed in practice, this could constitute 'permission.'

"It's the same with my tours—unless I encounter a particularly nasty official, park personnel just wave at me. I can't get formal permission, according to Parks Commissioner Adrian Benepe, who met with me when he was working at the NY Botanical Garden, because that could encourage lawsuits: People might try to sue the city, claiming they poisoned themselves by foraging, which the city allowed.

"This is all so irrational that I simply encourage people to do what makes sense: Forage safely and ecologically."

Gardens

Many of the "weeds" gardeners yank out and compost are great vegetables in their own right. A big percentage of my wild edible plants are ones that volunteered in my garden. I weed them out selectively, letting the choicest wild edibles have a spot right alongside the roses and oregano, but not letting them completely take over the garden. I've also introduced many species that are commonly found growing wild, including elderberry, Juneberry, spicebush, and mayapple. It's easier to harvest them in my own backyard, but more than that I use my now-domesticated specimens as indicators of when their wild cousins are likely coming into season.

If you have a garden, have a plot in a community garden, or know someone who has a garden, you can likely help yourself to the free food that would otherwise be composted.

Street Trees

There are mulberry trees, ginkgoes, basswoods, crabapples, and many other fantastic edibles planted as street trees all over New York City's five boroughs.

As far as I've been able to find out, if it's hanging over a public walkway it's legally fair game for anyone to pick. Many of the fruit trees were originally planted with the idea that neighborhood residents would collect the fruit. Now the fruits stain the sidewalk untasted while building owners curse them for being "messy."

Three blocks from me there is a white mulberry tree with especially sweet fruit (the quality can vary from tree to tree). When I saw a resident hanging out in front of the building, I asked if it would be all right with him if I harvested the fruit. He was delighted, since he'd been getting complaints about the "plant trash" on the sidewalk.

I go back every year now to get the mulberries from his tree. He e-mails me when they are ripe. It's a mutually beneficial relationship—a tasty one too.

"Empty" Lots

Leave any patch of earth empty for more than a week and wild plants will take it over, including many edibles. Most empty lots in the city boast at least half a dozen species of edible and medicinal plants. If there is an empty lot near you that isn't locked, go in and take a look.

Better yet, turn it into a community garden. That is how most community gardens get their start.

State and National Parks

I recommend checking online for each park's regulations before foraging in state and national parks. Part of the information I look for is what plants may be endangered in that particular park and therefore should not be harvested. Just because a plant is plentiful on one mountain doesn't mean that same species may not be in trouble a few counties over. This is especially true of some woodland plants, including such delicacies as ramps.

Perilous Suburbs

Foragers should be wary of suburban neighborhoods that boast perfectly manicured lawns and weed-free perennial borders. Many suburbanites still dump a hefty quantity of chemical fertilizers, pesticides, and herbicides on their gardens in order to get and maintain that tidy appearance. Although you might find people who are more than happy to let you collect the dandelions they haven't managed to eradicate from their lawn, a few questions may reveal that you really don't want to collect from such a toxic environment.

Foraging Sustainably

Here's a summary of basic foraging guidelines:
- **Harvest only where there is abundance:** If you find just one or a few of a species, don't harvest.
- **Never deplete an entire stand** of any species. Leave the biggest, healthiest plant to propagate future generations.
- **Never harvest more than 25 percent** of the leaves of an individual plant unless you know it is something so prolific that harvesting more won't harm the plant population. (Garlic mustard and mugwort come to mind.)
- **Never harvest a plant that is on the endangered** or threatened list for your area. (*Note:* Species that are plentiful in one state can be on the endangered list in another.)

Common Wild Edible Plants with No Poisonous Lookalikes

It's beyond the scope of this book to provide detailed plant identification information for individual wild edible plants (see Useful Resources for recommended book and online field guides), but here is a list of plants that are a good start for a fledgling forager. A quick search online (especially if you look up the italicized scientific name rather than the common name) will give you photos, drawings, and all the ID info you need to safely harvest these plants.

Amaranth (*Amaranthus* spp.)

Birch (*Betula* spp.)

Bramble berries (*Rubus* spp.)—blackberry, raspberry, dewberry, etc.

Burdock (*Arctium* spp.)

Cattail (*Typha* spp.)

Chickweed (*Stellaria media*)

Dandelion (*Taraxacum officinale*)

Hawthorn (*Crataegus* spp.)

Jerusalem artichoke (*Helianthus tuberosus*)

Lamb-quarter (*Chenopodium album*)

Mulberry (*Morus* spp.)

Nettle (*Urtica* spp.)

Oak (*Quercus* spp.)

Red clover (*Trifolium pratense*)

Rose (*Rosa* spp.)

Sassafras (*Sassafras albidum*)

Sheep sorrel (*Rumex acetosella*)

Violet (*Viola papilionaceae*)

Wild ginger (*Asarum* spp.)

Wood sorrel (*Oxalis montana*)

If You Do Just One Thing…

Feast on a wild plant you already know how to identify, the dandelion.

Wilted Dandelion Greens

Sometimes I've had a wild edible plants tour participant inform me that he or she tried dandelion greens and they were disgustingly bitter. "When was that? August?" I ask, because dandelion greens and many other wild greens are milder during cool weather than in summer.

The time to harvest dandelion greens in our area is in April before they flower. They still have a hint of bitterness then, but it is a hint only and actually adds to their tastiness.

MAKES 2–3 SIDE DISH SERVINGS

1 teaspoon oil or bacon fat
1 clove garlic, peeled and minced
1 large bunch early-spring dandelion greens, washed and torn or
   coarsely chopped
1 teaspoon vinegar
Salt to taste

1. Heat the oil in a frying pan over medium-high heat. Add the garlic and cook, stirring, for 30 seconds.

2. Add the dandelion greens and cook, stirring, until just wilted.

3. Sprinkle with the vinegar and salt to taste. Serve immediately.

Dandelion Root "Coffee"

When these dandelion roots are roasting in the oven, your home will be blessed with a tantalizing, molasses-like smell. Dandelion root "coffee" has no caffeine but tons of flavor.

1. Chop dandelion roots into pieces no thicker than ¼ inch and spread on a baking sheet.

2. Roast dandelion roots in a 300°F oven, stirring occasionally, until they are browned but not burnt.

3. Grind roasted roots in a coffee grinder. Brew as you would coffee.

Crustless Dandelion Quiche

MAKES ONE 9-INCH QUICHE, APPROXIMATELY 6 SERVINGS

2 teaspoons butter or oil

¼ cup dry bread crumbs

½ onion, finely diced

1 bunch dandelion greens, washed and chopped

1½ cups grated cheddar or other cheese (optional)

2 cups whole milk

3 large eggs plus one egg yolk, beaten

½ teaspoon salt

¼ teaspoon grated nutmeg (not local; optional)

1. Preheat oven to 350°F. Lightly grease a 9-inch pie pan with half of the butter or oil. Sprinkle with the bread crumbs. Shake off any excess.

2. Over medium-low heat, sauté the onion in a frying pan in the remaining butter or oil until translucent.

3. Wilt the dandelion greens in a large pot over medium-high heat stirring constantly. Add a splash of water if they start to stick. When wilted and cool enough to handle, squeeze any excess liquid out of the greens.

4. Spread the greens in the prepared pie pan. Top with the onion and the cheese, if using.

5. Wisk together the milk, eggs, salt, and nutmeg. Pour over the other ingredients.

6. Bake until the quiche is starting to brown on top in spots and a knife inserted into the center comes out clean, 30–40 minutes. Let stand 10 minutes before slicing, or let cool longer and serve at room temperature.

11

The Single Locavore

By the time I got seriously into local foods, I was divorced and living alone (well, with my cat). I did the entirety of my 250-mile diet year solo, so I am very familiar with the challenges and perks of being a single locavore.

"It is too much trouble to cook just for myself" and even "It's depressing to cook when there's no one to share it with" are objections I hear sometimes from my single friends. I respectfully disagree. Cooking a local foods meal for myself is a completely different experience from scarfing down some delivery pizza alone. Granted, it's also a different experience from sharing a meal with family or friends, but just as worthwhile.

Most of my meals alone are very simple ones. The ingredients are good and don't need a lot of fuss, and too often I'm tired from a long workday. Maybe I'll fix some steamed green beans with garlic and herbs and broiled chicken, or a greens-and-cheese-and-beans dish if it's a vegetarian night.

Because I know the people who grew my food (or I grew it myself and remember every week it took to get from seed to plate), I pay more attention to it than I might otherwise. If I'm making something with green beans, I remember that Farmer Ted told me that his workers find picking green beans so labor intensive that they'd rather do the weeding. A lot of work, care, and time went into these beans. Am I going to let them go to waste because "it's too much trouble to cook for myself"? That would be disrespectful to Ted, his workers, the beans, and myself.

Once in a while I cook fancier stuff for myself. Either I've got an idea for a recipe I want to try out or there's a cooking skill I want to master. When I have the time, I've been known to tackle solo soufflés, individual portions of baked local cheese with homemade chutney in puff pastry made with local flour. No, I'm not crazy. I have fun trying out those more complicated recipes. If they turn out beautifully, then I get excited and plan to serve them to the friends I have coming over for dinner next week. If the recipes flop, well, nobody knows except my cat and me.

Permission for Culinary Goofs Granted (Who's Going to Know?)

That freedom for audience-free culinary experimentation is one of the advantages of being a single locavore. While some creativity is a requirement for preparing meals with local foods (see chapter 8, Making Friends with Your Kitchen), if you're cooking for a family and your experiment with kohlrabi-plum casserole proves to be inedible, that could be a problem. But if you're on your own, you're free to spit it out, laugh, and take mental notes on what didn't work so that you don't make the same mistakes again. The experiments that turn out fabulously well get jotted down so that you can repeat and share them with dinner guests later on.

You might want to do more than take mental notes though. Since there may not be anyone to remind you about what worked so well with a particular recipe, unless your memory is a whole lot better than mine, you may not be able to re-create it later on. For some reason, what fails horribly is often much easier to remember than what worked gorgeously. I keep a kitchen journal where I write down the successes. Often these are not fully written out recipes but brief reminders, e.g., "grilled zucchini with tamari-honey dipping sauce—yes!"

Freeze Food in Single Portions

There's something discouraging about looking at an entire quart of soup frozen into a solid block when all I want is a small bowlful, or a pound of local chorizo when I just need 4 ounces for my recipe. I store almost all my frozen food in individual portions so that it is eater-friendly. The exceptions are fruits and chopped vegetables that I've frozen so that the pieces are loose and I can scoop out just what I need. (See "If You Do Just One Thing" in chapter 9, Simple Food Preservation.)

I Didn't Eat It Soon Enough and Had to Throw Food Away

We've all been there: You go to the farmers' market and the produce is so alluring with its bright colors and shapes and scents that you end up overstuffing your tote bag. On the way home you daydream about the virtuously healthy and delicious week you're going to have eating all those fruits and vegetables. But not tonight, because you have plans to meet friends for dinner. The next day you end up working three hours late, and . . . the strawberries are moldy, the lettuce is wilting, and now you're not feeling virtuous. You're feeling guilty for wasting both food and money.

One of the things about food that is picked when it is perfectly ripe and ready for harvesting is that it is perfectly ripe and ready for eating. Remember that this food was not grown or harvested with shelf life in mind. The time between perfect and spoiled can be short.

There are exceptions. Some local foods actually keep longer than their supermarket cousins. You'll find that locally grown greens like kale can last for weeks, especially if you store them in those green bags designed to keep produce fresh longer. (See Useful Resources for where to get them.) Hard vegetables such as kohlrabi and beets can last for a month or longer in the crisper drawers of your refrigerator. (See chapter 8, Making Friends with Your Kitchen, for more on the best ways to store your produce.)

The single locavore should think twice about buying an extra quart of juicy plums unless she is going to eat a lot of them and immediately freeze or otherwise preserve the rest. Fruit is top of the list of local foods that, if they're really ripe and wonderful, you have to eat or preserve within a day or two of bringing them home. The "use it or lose it" rule is especially true with very ripe strawberries and tomatoes (remember that tomatoes are technically fruits) because they should never be refrigerated. With other fruits you can buy yourself a day or two by refrigerating them.

All fruits freeze well, and so do most vegetables. If your strawberries are on the verge and you don't have time to make jam, put your freezer to work. See "If You Do Just One Thing" in chapter 9, Simple Food Preservation, for tips on the best way to freeze your produce.

Some vegetables that are, botanically speaking, fruits keep longer because they are deliberately harvested underripe. These include eggplants, cucumbers, and summer squash (a ripe zucchini looks like an oversized orange football, nothing like the version we are used to eating). These can keep in your refrigerator for a week or more.

Greens such as kale, mustard greens, bok choy, and chard keep longer cooked than raw. Also, the quality is better because they're cooked at their best rather than when they're on their way out. So if you've got a big bunch of greens that are way more than you can eat in one meal, go ahead and cook them all. Cooked, they'll last at least a week in the refrigerator. Don't season them yet. Season them as you use them so that you can vary the flavor from meal to meal. Sometimes my greens get garlic and olive oil; other days, local feta cheese and beans. Or they get blended with yogurt and herbs into dips, or added at the last minute to soups or to omelets, or . . . you get the idea.

The Whole Head of Lettuce and the Whole Loaf of Bread

Sometimes a whole head of lettuce is just too big for one person to use up while it's still fresh, even if you eat a salad every night. And unless you are majorly into breakfast toast and sandwiches, a whole loaf of bread can get stale and even moldy before you eat your way through it. Fortunately there are things you can do with lettuce besides salad, and things you can do with bread besides toast and sandwiches.

Braised Lettuce for One

This makes a lovely side dish that is ready in about 15 minutes.

MAKES APPROXIMATELY ONE CUP

1 teaspoon oil, butter, or bacon fat
1 small onion, chopped
Leaves from approximately half a head of mild, non-bitter lettuce,
    washed and torn in half
Salt or verdurette (See chapter 9, Simple Food Preservation, page 142.)
¼ cup peas, fresh or frozen (optional)

1. Heat the oil, butter, or fat over medium-low heat in a large skillet. Add the onion and cook, stirring occasionally, until soft and translucent.

2. Add the lettuce, and sprinkle with a pinch of salt or verdurette; add enough water to come halfway up the sides of the pan. Bring to a boil over high heat; reduce heat, cover pan, and simmer for 15 minutes. If using the peas, add them during the last 2 minutes of cooking.

3. Remove the lettuce from the liquid with a slotted spoon or tongs. Turn the heat to high and boil the liquid, uncovered, until it is reduced to a syrupy sauce.

4. Turn off the heat; return the lettuce to the pan and toss with the sauce. Add additional salt if desired. Serve warm.

Bread Crumbs

Bread crumbs are so easy to make that I can't believe anyone ever actually buys them. But I'll be honest—my food processor is what makes them easy to make.

Stale bread

1. Cut the crusts off the bread, and tear or cut into 1- to 2-inch chunks. Put in the food processor and process until the bread crumbs are the size you like.

2. Spread the bread crumbs on a baking sheet. Bake in a 200°F oven (I use my toaster oven), stirring often to keep them from burning or crusting over. You want them to dry out, not toast.

3. When they are dry and just beginning to turn golden, remove them from the oven and let cool. Despite the stirring while they were drying out in the oven, there will still be some clumps. I put mine back into the food processor at this point and give them a quick pulse to break up the clumps.

4. Put the crumbs into a container and store in the fridge. If you are adding new bread crumbs to old, shake the container to combine them. They keep pretty much forever.

Bread crumbs add a pleasantly crunchy topping to casseroles and grilled vegetables. Try tomato halves sprinkled with bread crumbs and broiled, adding torn basil leaves after they come out of the broiler. In traditional Italian cuisine, bread crumbs are often used on pasta in lieu of grated cheese. They can also be combined with a beaten egg and any finely chopped and already cooked vegetables plus cooked grains and/or meat to make meat loaf or veggie burgers.

Croutons

Use croutons for a crunchy texture play on salads and soups.

Stale bread

1. Cut the crust off the bread and cut the bread into ½- or 1-inch cubes.
2. Sprinkle with salt and drizzle with a little oil or melted butter and toss to distribute the salt and oil. Spread in a single layer on a baking sheet.
3. Bake in a 250°F oven, stirring occasionally, until lightly golden.
4. Let the croutons cool before storing in the refrigerator.

I Got How Many Cucumbers in My CSA Share?

Last week we got several foot-long Japanese cucumbers in our CSA shares. This week we got another three; next week we're supposed to get the same. Okay, people, I like cucumber salad and *tzatziki* (cucumber-yogurt dip) as much as the next person, but this is too much for a single locavore.

I made refrigerator pickles with some of them (see page 139). Most cucumber pickles are best with small, firm pickling cukes such as Kirby's, but refrigerator pickles are fine when made with oversize cucumbers as long as you scoop out the seeds before slicing them.

I also cooked some of them. Cucumbers are unexpectedly good cooked. They have a sweetness that is almost like snap peas. For cooked cucumbers, peel and seed a cucumber and cut it into ½-inch pieces. Cook in a frying pan over high heat with a little butter until they just turn translucent at the edges, about 2 minutes. Salt to taste.

Keep a Mental Note of What You've Got in the Fridge (or a Physical Note)

The fridge lists I described in chapter 7, The Convenience Factor, can be very helpful for the single locavore who doesn't want to throw food away because she forgot she had it until it had gone south. Although I do keep track of my preserved and frozen foods with actual lists, I'm not always that organized when it comes to the fresh produce that is constantly moving through my kitchen. But I do keep mental notes on what I've got and what of that is the most perishable. I also organize my crisper drawers into "Eat Soon" and "Will Keep." (See chapter 8, Making Friends with Your Kitchen, for more on how to do this.)

Take It Easy on Yourself

Once in a while, despite my best efforts, some of the produce in my refrigerator does end up in the compost bin. I feel a little better knowing that it's going to the compost rather than a landfill, but I still hate to "throw out" food. On the rare occasions when I do, I remind myself of something my single friend Kendall e-mailed me. Kendall—a local foods appreciator, if not a full-blown locavore—turns potential guilt into genuine celebration with this perspective:

"Sometimes when I buy too much greenmarket stuff, and may have to throw some away because it goes bad before I get to it, I think about the pleasure I had looking at the stuff on my counter, or even just in the greenmarket. I figure it's a bit like having cut flowers and freshness in the house. It's a visual pleasure and a reminder of a lovely day at the greenmarket, so no big deal if a few things don't get eaten."

Expecting Company

Having friends over for dinner is a joy whether it is an all-out party or just a quiet dinner for two. I used to stress over making the perfect meal for guests, but I've gotten over that.

I think the foodie magazines and television shows do us a disservice when they show us perfectly styled images of the "ultimate" recipe for this or that. It makes people try to cook at home as if they were restaurant chefs, but without the full staff of helpers to chop, stir, and do the dishes.

Cooking with friends, instead of cooking for them, can be a lot of fun. Sure it's gratifying to produce something impressive all on my own and bring it to the table with a big "Ta-da!" But nowadays when I want to try something complicated, I often let my guest or guests know that they are on sous-chef duty and ask them to come over a little early.

If cooking for guests had to be complicated, then I wouldn't do it too often. That would be a shame, because cooking for someone beside myself is one of life's great pleasures. Local foods make it easier to serve up something delicious without pulling out every pot and pan I own, because the ingredients are good when prepared simply. The food I cook for myself every day is perfectly acceptable company fare. And if the apartment isn't spotlessly clean because I didn't have time, so what? They're my friends, and they're coming over because we enjoy one another's company.

Other times, I do like to pull out all the stops. If I came up with an unexpected and spectacular flavor combination last week, company is my chance to share it with others. And if I recently taught myself how to make a complicated recipe, company is what does it justice. The extra time I put into these special-occasion foods is a way of letting my good friends know how much they matter to me.

Fancy or simple, sitting down together to share food is as much of a joy for the single locavore as for a family. As Kendall says:

"Invite people over to eat. I used to have biweekly afternoon meetings at my house, and I got the opportunity to experiment with lots of new (simple) vegetable dishes after shopping at the greenmarket. It is a great pleasure to make all this stuff, and people who would excuse themselves for nibbling on standard processed snacks or sweets are happy to gobble up lovely green things. They feel well cared for and enjoy the healthy, fresh food."

IF YOU DO JUST ONE THING...

Get a CSA half share.

A full community supported agriculture (CSA) farm share can be way too much food for one person unless you're seriously into food preservation. Here are a few of the options some CSAs offer to single locavores:

1. **Get an every-other-week (EOW) share**, where you pick up a full farm share but only every other week. That could still be more food than you can eat before it spoils if you're not doing a little preserving, but it's a good option if you are.

2. **Share a share.** You can team up with a friend for a weekly (or EOW) share and split the bounty. Some CSAs will help pair you with someone who wants to share a share if you don't have an eager candidate. This is a good option if you only want an amount of fresh food that you can consume in a week, but it can get tricky.

   Let's say you met your share mate at the CSA distribution after work, and this week the share includes a whole pumpkin and a head of lettuce. Since you and the person you're sharing with don't normally carry kitchen knives in your briefcases, how are you going to split up that pumpkin and head of lettuce?

   Some solutions that people have come up with at my CSA include deciding who likes what better—"I'll take the pumpkin, you take the lettuce"—or scheduling weekly kitchen time together on the night of a CSA distribution. You go over to your friend's house or vice versa, enjoy conversation while cooking dinner, and while you're at it carve up that pumpkin and divvy up the lettuce leaves so that you can share them.

3. **Get weekly half shares.** Some CSAs offer weekly half shares. You pick up every week but only take half of what the full-share members are getting. Single items like the pumpkin and head lettuce mentioned above are offered as either/or options to half-share members. For example, you choose whether you want the pumpkin or the lettuce this week, but only full-share members take both.

12
The Space-Challenged Locavore

When Martha Stewart told me that I really must get a root cellar, she clearly had never seen my apartment. It's a typical small New York City one-bedroom, except for the fact that I am lucky enough to share a garden with the neighbors in the apartment next door. I can cross my kitchen, which is also my living room and dining room, in four strides. The hallway is too narrow to put in shelves. Storage space is minimal, counter space almost nonexistent.

None of this would be a problem if I weren't a locavore. Extra stuff can always be put into a storage unit, as many urbanites (including myself) do. If you're not cooking with fresh ingredients, you don't need much in the way of kitchen equipment or counter space. But what if you're not only cooking fresh food but also putting up food for the winter?

Even people whose abodes are much more spacious than mine often lack space devoted to food. I know one family on the Upper West Side with a spectacular apartment. It has an amazing view of Central Park, several bedrooms, four bathrooms . . . and a kitchen the size of a closet. Clearly the architect was envisioning restaurant dining and delivery food, not home cooking.

I am fascinated by old cookbooks and the glimpses they give me of daily life in other eras. "The crock may be kept in a cool cellar until fermentation ceases," instructs one recipe for sauerkraut. Cool cellar? What, you mean my refrigerator?

In many urban homes, the refrigerator has replaced the cellar. Kitchen cupboards have replaced pantries, often with just enough room to store dishes, forget about food. Counter space is minimal. Few of my friends' kitchens can boast room for a kitchen table.

Here are some ways I manage a locavore's life in a space-challenged apartment.

Lose the Clutter

When I lived in a much bigger apartment, I accumulated stuff—stuff that apparently I didn't really need. When I moved into smaller digs, a lot of that stuff went into boxes in a storage unit. Guess what? Six years later I still haven't opened those boxes, and I don't actually remember what is in them anymore. I'm still deciding when to have the stoop sale and get rid of whatever is in there.

Choreographer Francis Patrelle says renting a storage unit is the ultimate optimistic act because it implies that you expect to have a bigger place someday. That's swell, but until you move into your dream lair, a storage unit won't help you find counter space while you're cooking, or a place to store your home-canned tomatoes. (The contracts at storage facilities specifically prohibit food storage.)

With the exception of those mystery boxes still in storage, I've learned to get rid of whatever I don't cherish or need on a daily basis. I rely on the unofficial Park Slope recycling service: Items I put out on the street in front of my building—clothes, books, whatever—disappear within five minutes, claimed by unknown neighbors. That's also how I found much of my furniture, but never mind.

The point is, get rid of the stuff you don't use or love having around. Put it in storage if you can't bear to part with it (maybe we can have our stoop sale together). Then get creative with how you use the space you do have.

Everything Is Counter Space

One of the major complaints about tiny kitchens is that they lack counter space for chopping, setting bowls on, etc. In my kitchen, everything is a potential work surface.

For example, I have two wooden bar stools that nobody ever sits in unless I'm having a party. One is placed right next to the stove, the other beside the table where I do most of my chopping and slicing. When I need to set aside a mixing bowl of something to make room for a cutting board, onto one of the stools it goes.

If I'm really having a domestic goddess cooking frenzy, my two dining room/kitchen chairs also become resting places for pots, pans, and bowls.

And don't forget the kitchen sink: If you aren't actually using it to wash anything and there aren't any dishes in it, why shouldn't you put that skillet of sizzling local chorizo there for a minute while you put a different pan on the stove?

Everywhere Is Storage Space

Forget the use-labels things come with. An oven doesn't have to be just for baking and roasting, a bookshelf doesn't have to be just for books.

Inside other things is useful storage space. I keep my frying pans and baking sheets in the oven, for example. I just have to remember to take them out before preheating to bake something.

Remember that there's no rule that says food, nonperishable food at any rate, has to live in the kitchen. Jars of my home-dried mushrooms and celery share bookshelf space with books in my bedroom.

Obviously, moisture-sensitive things like books shouldn't be next to moisture-laden perishables such as berries. My fresh produce goes on whatever available surface in the kitchen or in the refrigerator. (See chapter 8, Making Friends with Your Kitchen, for which goes where.)

Look Under

If you have cabinets over your kitchen sink, and space under those cabinets, screw some hooks into their undersides. Now you've got a great place to hang your coffee and tea mugs without using up shelf space.

And since I mentioned my bedroom, don't forget that there is space under the bed. (See "If You Do Just One Thing" below.) Under the sofa might be an option as well. That is where a lot of my empty-but-soon-to-be-filled canning jars live.

On the Wall

If your building is all right with putting nails, hooks, or screws into the walls, then you've got lots of vertical storage space. Almost all my pots and pans are hanging on the wall, freeing up my limited cabinet and counter space. I've also attached a canister to the wall that holds my spatula, cooking spoons, and whisks.

On the Sides of the Refrigerator

Some of my spices and all my wooden spoons and graters are in little metal containers that have magnets. They work great clinging to the metallic sides of my refrigerator.

Look Up

Besides looking inside the oven and under my furniture for kitchen equipment and nonperishable food, look up. Some brilliant tenant who lived in my apartment before me put in a wide overhead shelf that runs the length of my hallway. As I mentioned, my hallway is too narrow to put in shelves at floor level—I wouldn't be able to squeeze by. But that overhead space is out of the way. Obviously it's also a little bit of a pain to get to; I have to get the stepladder out for that. So I use the shelf to store infrequently used cookbooks and other things that I want available but don't actually need to get access to every day. That frees up space on the easier-to-reach shelves in the rest of the apartment.

Also in the "look up" category, I use the space on top of my kitchen cabinets for storing stuff I use only occasionally. I've got serving dishes I don't use very often up there, extra wine glasses for parties, empty canning jars. I don't store canning jars filled with food that close to the ceiling, because heat rises and it gets a little too hot up there for food storage. Also, that area gets a lot of light in my apartment, and most canned goods should be stored away from direct light.

Everything Is Storage Space, Except . . .

Don't store any food on top of the refrigerator except for ingredients that are already very dry, such as dried pasta and crackers. The temperature on the top of the refrigerator is extremely warm. Nuts and grains, including grain flours, will go rancid if stored there. Dry beans, which are not actually completely dry but contain enough internal moisture to support the embryo of a bean plant, become uncookable if stored in a hot place like the top of the refrigerator for too long.

I found that out the hard way. Local beans from Cayuga Organics that had been so fresh when I first got them that they didn't need the precooking soak beans usually do turn into pebbles after a few months on top of the refrigerator. I soaked them and simmered them for days, and they were still inedible. In general, the top of the refrigerator is only good for storing extra kitchen gear.

Making the Most of Shelf Space

Because the few kitchen cabinet shelves I have are packed to the max with jars of food and spices, it could be hard to know what's in the back of the shelf. If I wanted the dried thyme and there were three other rows of jars in front of it, I would have to take out all the front jars just to find the one I wanted. My solution to this is that I've put in small lazy Susans on my shelves. These spin around so that I can get at the things at the back of the shelves easily. You can buy them cheaply at hardware and household goods stores.

If you have pullout kitchen drawers with some depth to them, they can be good for storing small jars of spices and herbs. Just be sure to label the lids so that you don't have to take each jar out to see what it is.

Edible Decor

There are people who pay good money to decorate their homes "country" style. They have big bottles of colorful pickled vegetables that no one is ever going to eat, layers of interesting looking beans in jars no one is ever going to open, attractively crafted swags of dried herbs gathering dust on the walls.

I have all that decor—but with the important difference that I eat it.

Once I had a houseguest visiting from Australia. While I cooked our dinner he commented on how charming and quaint my apartment was. When I snapped several sage leaves off a pretty bunch pinned to the spice rack and threw them into our soup, he looked shocked. I guess it had never occurred to him that the herbs were there for anything more than decoration.

Putting the attractive stuff out where you can see it frees up closed cabinet space for other things. (I don't think anyone could mistake my jars of home-canned chicken stock for decoration if I put them out.)

Know Thy Leftovers

This bit of excellent advice comes from Kendall, who lives in a small apartment in the West Village:

"Get to know which leftovers you will eat and which just get thrown away. For example, I will eat leftover chicken for sure—but leftover steamed kale often gets neglected and I have to throw it away. So now I try to know which things I can cook in excess for leftovers and which I should just cook for one meal."

I do use leftover cooked greens, including kale, but there are other things I don't get around to. "Know thyself" is a good motto to follow in the kitchen.

Kendall's advice also applies to preserves. One year I made jam out of every single fruit I could get hold of as it came into season. I loved the names, the colors, the way the jars looked lined up on my shelves. Juneberry jam, red currant with strawberry, yellow plum, elderberry . . . I loved the way each one tasted too.

But there was a problem: I don't eat much jam. Breakfast toast is an occasional thing for me, and although I got creative finding ways to eat my jams (jam with yogurt, apricot jam glaze for meat, etc.) and gave lots of jars away as gifts, I still had dozens of jars taking up space, space that I could better have used for foods I eat more often. Lesson learned. I still make a few jams for myself and for gifts, but most of my preserves now are the things that I will actually eat up.

Clean As You Go

When you're cooking in a teeny kitchen, you need every work surface you've got, and there just isn't anyplace to pile up dirty pots and pans and measuring cups. If you have a dishwasher you can start stacking in there as you go. I don't have one, so I wash as I go. This doesn't take extra time—I do it while something is simmering on the stove or the bread is baking in the oven. And there's a benefit besides keeping workspace available: At the end of the meal, there's very little cleanup left to do.

If You Do Just One Thing . . .

The Under-the-Bed Solution

Who says food and kitchen gear have to be kept in the kitchen? It's actually a luxury to have a separate room devoted to food and cooking. In generations past (and in many less-developed nations today) many homes were just a single room.

Think of your home as your own personal urban homestead, a sort of log cabin with edibles stashed wherever there's room and even hanging from the rafters (okay, not literally rafters, but you know what I mean). Food and cooking gear can live wherever you have room.

One of the places there is room is under your bed or sofa. Get yourself some storage bins that are shallow enough to fit under the bed but deep enough to hold pint jars. The kind that have sliding drawers are best because you don't need to pull out the whole bin and take the entire lid off every time you want a jar of something. Be sure that whatever you store there is labeled on top of the lids so that you can see what you've got at a glance.

Anything nonperishable can be stored this way, including sealed jars of canned food, dehydrated food, and kitchen gear you don't need every day.

183

Afterword
The Hundredth Monkey Effect

Recently I read that if conventional food supply lines were cut off, New York City would have only forty-eight hours worth of food.

If that happened, I'd be turning to my garden, the parks, my pantry of home preserves, and maybe the local farm trucks if they could still get through to deliver to the farmers' markets and CSAs.

If push really comes to shove, personal food security isn't cross-country trucking routes or international trade agreements. It's a garden or a direct connection to a nearby farm or some foraging know-how.

But the old guard is mighty, and they aren't going to change willingly. The conventional agriculture corporations have launched numerous attempts to discredit the local and organic food movement. When Michelle Obama announced that her White House vegetable garden was going to be an organic one, the Mid America CropLife Association (MACA), which represents companies selling chemical pesticides and fertilizers, sent Mrs. Obama a letter suggesting that she should reconsider her decision to garden organically. They urged her to use their chemicals, which they euphemistically referred to as "crop protection products." An e-mail to MACA's supporters from a spokesperson read, "While a garden is a great idea, the thought of it being organic made Janet Braun, CropLife Ambassador Coordinator, and I shudder."

It's not just the corporate agribiz folks who are still miles away from even considering the changes needed to create a sustainable food system. One of the neighbors I share my garden with is a college kid who freely admits he doesn't cook and knows "I should think more about where my food comes from, but that's pretty far away for me at this point." I've seen him sitting out in the garden, surrounded by fruits, herbs, and vegetables, eating from a Styrofoam take-out container.

I'm not discouraged. Although agricultural change is urgent for our health and the health of our planet, not everyone has to change all at once. Small changes do make a difference, and the numbers of us making those changes is growing.

All we need is one hundred monkeys.

In the early 1980s Lyall Watson reported that in the 1950s Japanese primatologists studying macaque monkeys in the wild had observed a fascinating phenomenon. On the island of Koshima, scientists were providing the monkeys with sweet potatoes dropped in the sand. The monkeys liked the sweet potatoes but not the sand that clung to them.

An eighteen-month-old female figured out that she could solve the gritty problem by washing the potatoes in a nearby stream. Her mother copied the trick, and soon her playmates and their mothers did too. Over a period of six years, the scientists watched the younger monkeys all figure out how to wash the sand off the sweet potatoes. Some of the adults followed suit, while others kept eating the gritty potatoes.

But once a certain number of monkeys had adopted the new method, a tipping point was reached. Within a short time almost all the monkeys on the island were washing their sweet potatoes.

Ken Keyes, writing about the same subject, dubbed the phenomenon "The Hundredth Monkey Effect." Although the exact number of monkeys that added up to the tipping point is unknown, Keyes chose one hundred as the number to describe what he thought had happened. Once the critical number was reached, not only did the other monkeys on Koshima pick up the new habit at a much faster rate than they'd been doing over the six years since the first monkey figured it out, but something even more astonishing happened.

The scientists observed that monkeys on other islands, as well as a mainland group of monkeys at Takasakiyama, began washing their sweet potatoes too.

It may not be necessary for the majority of people to change their food habits in order for substantial change to occur. We just need the minority to keep growing until it achieves the hundredth monkey effect.

Whether your main concern is the health of yourself and your family, climate change, your local economy and small farms, food safety and security, clean waters and healthy soil, or simply exquisite food, I'm encouraged that you've read this book and are thinking about these things.

Monkey Ninety-eight, meet Monkey Ninety-nine . . .

185

Useful Resources

The Web sites and books below will connect you with organizations, individuals, and information that provide practical resources for eating local. Some of them also provide inspiration through the experiences and stories of your fellow locavores, and all of them connect you to the rapidly growing community of the local food movement.

1: My Year as a Locavore
Books and blogs by locavores who have done 100-mile and 250-mile diets.

Animal, Vegetable, Miracle
www.animalvegetablemiracle.com
Book and Web site by Barbara Kingsolver, Camille Kingsolver, and Steven L. Hopp.

Botany, Ballet & Dinner from Scratch: A Memoir with Recipes
Leda Meredith tells the full story of The 250 in her book.

Coming Home to Eat: The Pleasures and Politics of Local Food
Gary Paul Nabhan did a 250-mile diet for a year and chronicled his experiences long before local food was a hot topic. He writes eloquently and informatively about his experiences in this book.

Leda's Urban Homestead
http://ledameredith.net/wordpress
The blog that started with The 250 and is still where I report on my local food, foraging, and gardening adventures. Along with my latest local food finds, adventures, successes, and mistakes, you'll find recipes, tips, and updates on local food-related classes, including vegetable and herb gardening and foraging.

The Omnivore's Dilemma: A Natural History of Four Meals
Michael Pollan
This is the best-selling book that gave the local food movement momentum and got thousands of people to consider where their food is coming from.

Plenty: One Man, One Woman, and a Raucous Year of Eating Locally
www.100milediet.org
Book and blog by Alisa Smith and J. B. Mackinnon.

2: How Can Eating Great Food Save the World?

Web sites, books, and films on how your food choices make a difference.

Cool Foods Campaign

http://coolfoodscampaign.org

The Cool Foods Campaign educates the public about how food choices can affect global warming and empowers them with the resources to reduce this impact.

Eat Here: Reclaiming Homegrown Pleasures in a Global Supermarket

Brian Halweil's excellent book on the importance of local food and the options available to today's locavores.

Farm Aid

www.farmaid.org

Organization working to keep family farmers on their land.

Food Democracy Now!

www.fooddemocracynow.org

A grassroots movement initiated by farmers, writers, chefs, eaters, and policy advocates who recognize the profound sense of urgency in creating a new food system.

Food, Inc.

www.foodincmovie.com

Documentary film by Robert Kenner on the industrial food system.

The Future of Food

www.thefutureoffood.com

Documentary film by Deborah Koons Garcia that covers the impact of genetically modified food on farmers, consumers, and the environment.

In Defense of Food: An Eater's Manifesto

Michael Pollan's book urges us to return to eating food our great-grandmothers would recognize as food and explains why doing so matters to our health and happiness.

King Corn

Documentary film by Ian Cheney and Curt Ellis on the impact of corn subsidies and farm policies on our food and environment.

Local Harvest

www.localharvest.org

Local Harvest's site can help you locate CSAs and farmers' markets nationwide.

The Meatrix
www.themeatrix.com
Animated video about how the meat you eat is really raised and how to locate better options.

Nourishing Traditions: The Cookbook That Challenges Politically Correct Nutrition and the Diet
    Dictocrats
Sally Fallon explains why low-fat and vegan diets may not be as healthy as we've been led to believe.
    Her book also includes many excellent and unusual recipes.

The Revolution Will Not Be Microwaved: Inside America's Underground Food Movements
Sandor Ellix Katz takes an inspiring look at some of the individuals, including farmers, who are
    finding alternatives to the industrial agriculture food chain.

Stuffed and Starved
www.stuffedandstarved.org
Blog by author Raj Patel who writes on food, hunger, and globalization.

Sustainable Table
http://sustainabletable.org
Info and resources about the sustainable food movement.

Worldwatch Institute: Vision for a Sustainable World
www.worldwatch.org
Includes several excellent sustainable agriculture blogs.

3: Sourcing Local Food
NYC–area and national resources for finding farmers' markets, CSAs, and other sources of local
foods near you.

Alternative Farming Systems Information Center
www.nal.usda.gov/afsic/pubs/csa/csa.shtml
This part of the USDA's Web site includes information on CSAs and links to sites that will help you
    locate one near you.

BK Farmyards
http://bkfarmyards.com
If you've got sunny garden space but lack the time or gardening know-how to tend the vegetable
    garden of your dreams, BK Farmyards will do it for you.

Just Food
www.justfood.org
Comprehensive resources for finding, starting, or managing a CSA in New York City.

Local Fork

www.localfork.com

An organization in NYC that makes it easier to eat local by providing the tools that stimulate grassroots local food networks.

Local Harvest

www.localharvest.org

Local Harvest's site can help you locate CSAs and farmers' markets nationwide.

The Locavore's Guide to New York City

www.localfork.com/locavoreguidenyc.aspx

A directory organized by product to help New York City residents find locally produced ingredients. This is a project I started with Local Fork.

Pick Your Own

www.pickyourown.org

Locate a U-Pick farm near you.

4: Eating with the Seasons

Online resources for finding out more about what is in season when.

Field to Plate

www.fieldtoplate.com/guide.php

Online directory of sites devoted to what food is in season when for each U.S. state.

Natural Resources Defense Council's Food Page

www.nrdc.org/health/foodmiles/

Another online directory that lists what is in season when in the region where you live.

5: The Zero Miles Diet: Grow It Yourself

Books, Web sites, and NYC–area classes for the home gardener.

Brooklyn Botanic Garden

www.bbg.org

BBG offers classes for all levels of gardeners, many of them focused on growing vegetables and herbs. Topics include victory gardens, edible weeds, and container gardening.

Brooklyn Greenbridge

www.bbg.org/edu/greenbridge

Works with block associations, community gardens, and community centers to promote conservation and community through gardening activities.

Bronx Green Up

www.nybg.org/bronx_greenup

Offers horticultural advice, technical assistance, and training to community gardens, school groups, and other organizations interested in improving urban neighborhoods in the Bronx.

Companion Plants

www.companionplants.com

Good mail-order source of a large collection of high-quality herb plants.

Down and Dirty Gardening

www.downanddirtygardening.com

Local foods enthusiast Ellen Zachos's blog with gardening, foraging, and cooking tips.

Fungi Perfecti

www.fungi.com

Excellent source of growing kits for gourmet mushrooms.

Green Guerillas

www.greenguerillas.org

Uses a unique mix of education, organizing, and advocacy to help people cultivate community gardens, sustain grassroots groups and coalitions, engage youth, paint colorful murals, and address issues critical to the future of their gardens.

Green Thumb

www.greenthumbnyc.org

The largest community gardening program in the United States, supporting over 600 member gardens in New York City. Their Web site includes a link to help you find a community garden near you. They also have resources to help you turn that abandoned lot in your neighborhood into a garden.

Hudson Valley Seed Library

www.SeedLibrary.org

Offers seeds for heirloom vegetable varieties that were specifically bred for the Northeast.

New York Botanical Garden

www.nybg.org

NYBG offers demonstrations and classes related to food during its summer-long edible garden event, as well as ongoing classes and a certificate program for gardeners.

NYC Compost Project

www.nyccompost.org

Discounted compost bins, worm recycling kits, classes and information on composting.

Pinetree Gardens

www.superseeds.com

Pinetree sells smaller amounts of seeds in their packets and offers correspondingly lower prices. This is a plus for the home gardener, because who really needs 200 zucchini plants in their backyard? (That's a typical amount of seeds per commercial packet.) Pinetree carries interesting vegetable varieties from around the world and a good selection of herb seeds as well.

Plants for a Future

www.pfaf.org

U.K. site that hosts a large database of information on edible and otherwise useful plants, including many wild ones. Includes tips on landscaping with edible plants.

Seeds of Change

www.seedsofchange.com

Large selection of organic vegetable and herb seeds. Smaller but still good selection of vegetables and herbs sold as mail-order plants rather than seeds.

Self-Watering Planter

www.josho.com/gardening.htm

Detailed instructions on how to make your own earth box, aka self-watering planter.

Silver Heights Farm

www.silverheightsfarm.com

Wonderful selection of heirloom vegetable and herb plants. If you live in New York City, you can e-mail Silver Heights orders for pickup at the Union Square Greenmarket.

United Plant Savers

http://unitedplantsavers.org

An organization that works to protect native medicinal plants of the United States and Canada and their native habitat while ensuring an abundant renewable supply of medicinal plants for generations to come. You can order bare-root edible and medicinal woodland plants from them.

6: The Cost Factor

Web sites and organizations working to bring local food to low-income families and individuals.

City Harvest

www.cityharvest.org

Rescues food that would otherwise go to waste for food pantries and soup kitchens, including the leftovers from CSA distributions and farmers' markets.

Co-op Directory Service

www.coopdirectory.org

Locate a food co-op near you.

Food Systems NYC

www.foodsystemsnyc.org

An organization working to increase access for all to safe and wholesome food, and to strengthen and expand the regional farm and food economy.

7: The Convenience Factor

Products, directories, and a phone app to make your locavore lifestyle easier.

City Distance Tool

www.geobytes.com/citydistancetool.htm

Enter where you are and the name of the nearest town to where a particular farm is located and this online service will tell you how many miles from farm to your plate.

Kool-to-Go Bags

www.kooltogo.com

Source for thermal bags that will keep your food cool or even frozen for hours if you can't get home from the farmers' market right away.

Locavore App

A handy app for iPhone users that shows you what's in season and what's about to be in season; has a full listing and calendar of farmers' markets in the NYC area.

The Locavore's Guide to New York City

www.localfork.com/locavoreguidenyc.aspx

A directory organized by product to help New York City residents find locally produced ingredients. This is a project I started with Local Fork.

8: Making Friends with Your Kitchen

The cookbooks I use most often.

Botany, Ballet & Dinner from Scratch: A Memoir with Recipes

Leda Meredith includes the full story of The 250 and many recipes and food preservation tips.

Chez Panisse Vegetables

Alice Waters, one of the gurus of the local, organic food movement, wrote this wonderful cookbook. Simple, lovely recipes organized by vegetable—perfect for when you get bok choy in your CSA share five weeks in a row and can't think what else to do with it.

Chez Panisse Fruits
Another Alice Waters classic includes some food preservation recipes.

Joy of Cooking
Rombauer, Becker, and Becker. The cookbook to own if you're only going to have one.

Wild Fermentation: The Flavor, Nutrition, and Craft of Live Culture Foods
Sandor Ellix Katz's book on everything fermented, including lacto-fermented vegetables, sourdough
    baking, yogurt, and more.

9: Simple Food Preservation
My favorite books on food preservation, plus online sources for canning supplies and information.
Also, a winter CSA that includes preserved foods.

Fruits of the Earth
Just when I thought I didn't need one more cookbook or food preservation book on my limited shelf
    space, I checked this one by Gloria Nicol out of the library. By the time I'd bookmarked more
    than a dozen of the inventive recipes, I gave in and bought a copy.

Home Canning Supply
http://homecanningsupply.com/
Links to many sources of inexpensive canning supplies.

National Center for Home Food Preservation
www.uga.edu/nchfp
Great online instructions for different food preservation methods.

***Nourishing Traditions: The Cookbook That Challenges Politically Correct Nutrition and the Diet
    Dictocrats***
Sally Fallon offers excellent recipes for fermented foods including yogurt and lacto-fermented fruits
    and vegetables.

Preserving Food without Freezing or Canning
Written by the gardeners and farmers of Terre Vivant, it includes traditional techniques and recipes
    for food preservation that require no special equipment. One of my favorite food preservation
    books.

Putting Food By
Janet Greene, Ruth Hertzberg, and Beatrice Vaughan
This is the bible as far as understanding what makes different food preservations safe (or unsafe).

Small Batch Preserving
Ellie Topp and Margaret Howard
Good basic preserving recipes that yield only a few jars each. Perfect for small kitchens.

Vinegar Making & Testing Instructions

www.naturemoms.com/homemade-vinegar.html

Recipes for homemade vinegar and how to test yours to find out if it is acidic enough for safe food
preservation.

Winter Sun Farms

www.WinterSunFarms.com

Winter Sun Farms offers a winter CSA that includes frozen berries, greens, tomatoes, and other
vegetables.

10: Feasting for Free
Books and Web sites devoted to wild edible plants and mushrooms.

The Audubon Guide to Mushrooms

Edited by one of my favorite instructors, expert mycologist Gary Lincoff.

The Eyewitness Guide to Mushrooms

Another book edited by Gary Lincoff, this one relies more on photographs than the *Audubon Guide*
does.

Forage Ahead

http://tech.groups.yahoo.com/group/ForageAhead/

Wild edible plants enthusiast e-mail group for both novice and experienced foragers.

Forager's Harvest

www.foragersharvest.com

Sam Thayer's terrific wild edibles site; the best place to get his book (one of my favorite and most
trustworthy guides to wild foods).

Foraging with the "Wildman"

http://wildmanstevebrill.com/

"Wildman" Steve Brill's excellent wild edible plants and mushrooms site; the best place to order his
wild edibles field guides and cookbooks.

Wild Food Adventures

www.wildfoodadventures.com

Wild edible plants Web site by John Kallas, PhD. Although he is on the West Coast, most of the
information here is also excellent for Northeast foragers.

11: The Single Locavore
Great single-serving cookbooks.

Going Solo in the Kitchen

Jane Doerfer. One of my favorite books on cooking for one.

Serves One: Simple Meals to Savor When You're on Your Own
Toni Lydecker. Good recipe inspirations for one.

Solo Suppers: Simple Delicious Meals to Cook for Yourself
Joyce Goldstein. Excellent single-serving recipes that can easily be adapted to local ingredients.

12: The Space-Challenged Locavore
A helpful book for locavores with small homes.

500 Ideas for Small Spaces: Easy Solutions for Living in 1000 Square Feet or Less
Kimberly Seldon. Fun solutions for dealing with lack of space.

Index

200

201

About the Author

Leda Meredith has become New York City's "locavore emeritus." She is the author of *Botany, Ballet and Dinner from Scratch,* a memoir with recipes about her year of eating only foods grown within a 250-mile radius of her home in Brooklyn, New York, and has been featured on *The Martha Stewart Show,* in the *New York Times* and *Post,* and on locavore blogs and Web sites. She is an instructor at the New York Botanical Garden and the Brooklyn Botanic Garden specializing in edible and medicinal plants. In addition to teaching ongoing classes, Leda regularly leads foraging tours, and she speaks and  leads seminars at Brooklyn Botanic Garden, New York Botanical Garden, Genesis Farm, Stone Barns Institute, Woodbridge Garden Club, and The Herb Society of America.

Leda details her locavore lifestyle on her blog, Leda's Urban Homestead: ledameredith.net/wordpress.